Edlef Köppen
Heeresbericht

RAPID FIRE

Report of the Army Command

First published in 1930 by Horen Verlag Berlin

Edition 2 2025
Translation by Sven Dethlefs

Cover Concept by Leni Dethlefs and Anne Kathrin Wille
Editing, Proofreading, and Cover Design by ebooklaunch.com
Formatting by Andrea Reider
Cover Image: *Illustration for Covenants with Death* (Daily Express Publications, 1934), Bridgeman Images
Afterword using publications by Jens Malte Fischer (DVA 2004) and Wilhelm Ziehr

Note by Edlef Köppen (Horen Verlag 1930):
The names of people and military units do not correspond to reality except for the ones mentioned in the contemporary documents.

ISBN 979-8-88170-218-2

This translation is dedicated to the memory
of my grandmother Betty,
who lost her brothers Fritz and Werner
in the World Wars

Western Front

1914: *First Battle of the Marne* – German advance halted

1915: *Battle of Ypres* – Introduction of poison gas

1916: *Battle of the Somme* – Failed Allied offensive with the first use of tanks in warfare

1916: *Battle of Verdun* – One of the longest and bloodiest battles, primarily a French-German conflict

1917: *Battle of Passchendaele* – Allied forces push back German lines but suffer heavy losses

1917: *Battle of the Lorette Heights* – French forces capture the strategic heights, weakening the German defensive line

1918: *Ludendorff Offensive* – A final German attempt to break the stalemate; initial success but ultimately failed

1918: *Hundred Days Offensive* – Allied counteroffensive, leading to the German retreat and armistice in November 1918

Eastern Front

1914: *Battle of Tannenberg* – Decisive German victory over invading Russian forces in East Prussia

1915: *Brusilov Offensive* – Major Russian offensive, initially successful but ultimately slowed by German reinforcements

1916: *Battle of Lake Naroch* – Russian attempt to relieve pressure on Verdun but failed to make significant gains

1917: *Russian Revolution* – Russian collapse and withdrawal from the front due to instability following Lenin's revolution

1918: *Treaty of Brest-Litovsk* – Russia signs peace treaty with Germany, leaving the Eastern Front under German control

Notes on This Translation

Köppen uses military ranks specific to the Prussian Field Army and the field artillery branch. While officer ranks such as *Leutnant*, *Hauptmann*, and *Major* correspond directly to their British Army counterparts, German non-commissioned officer (NCO) ranks lack exact equivalents in British or American military hierarchies.

The term *Unteroffizier* refers both to the entire NCO class and to the entry-level NCO rank. Köppen alternates between using this general designation and specific NCO ranks. In this translation, "sergeant" is generally used to represent NCOs, with specific rank titles introduced only when clarification of the chain of command among NCOs is required.

In direct speech, this translation retains the original German military ranks and the honorific *Herr* to reflect the formal, hierarchical mode of address integral to Prussian military tradition. In German, *Herr*—comparable to "sir" in English—is a standard form of polite address used for someone older, of higher status, or outside one's close circle. In Köppen's depiction, soldiers always address superiors using *Herr* followed by the rank, while superiors address subordinates by rank alone, omitting *Herr*.

Translator's footnotes have been added for context and explanation. They are not in the German edition.

In his book, Köppen switches between the present tense and the past tense, sometimes in the same chapter. This translation takes over the German original as it is not clear whether Köppen deliberately changed the tenses or whether it was an oversight by his editor.

Military Ranks

- Major – Major; commands a brigade or battalion
- Hauptmann – Captain; usually commands a company or battery
- Oberleutnant – First Lieutenant; a captain's deputy
- Leutnant – Lieutenant; entry-level officer rank
- Ordonnanzoffizier – Ordnance Officer; a staff officer who assisted a commanding officer, often serving as an "empowered messenger," delivering and, if required, explaining the tactical instructions of the commanding officer
- Unteroffizier – General term for NCOs or the entry-level NCO rank
- Wachtmeister – Master Sergeant; highest NCO rank in field artillery
- Vize-Wachtmeister – Staff Sergeant; one rank below Wachtmeister
- Gefreiter – Private First Class or Lance Corporal

Edlef Köppen
Heeresbericht

RAPID FIRE

Report of the Army Command

Translation by Sven Dethlefs

To those
who kept me alive

Chief Censorship Office No 123 O.Z.

March 23, 1915

It is not desirable for the reports covering major periods of the war to be published by people who, because of their position and experience, may not have been able to grasp the entire context properly. The emergence of such a type of literature would lead to a very one-sided assessment of the events in broad parts of the population.

Part I

Chapter I

1

We, Wilhelm, by the grace of God German Emperor, King of Prussia, etc., decree the following on the basis of Article 68 of the Constitution of the German Reich: The territory of the Reich, with the exception of the royal Bavarian territories, is hereby declared to be in a state of war. This Ordinance shall enter into force on the day of its promulgation. Authenticated under our most personal signature and with the Imperial Seal affixed.

Given Potsdam, Neues Palais, July 31, 1914

Wilhelm I.R.
von Bethmann Hollweg[1]

Mobilization

I hereby decree:

The German Army and the Imperial Navy are to be prepared for war in accordance with the mobilization plan for the German Army and the Imperial Navy.
 August 2, 1914, is set as the first mobilization day.

Berlin, August 1, 1914.
Wilhelm I.R.
von Bethmann Hollweg

---

[1] Theobald von Bethmann-Hollweg, 1856–1921. Chancellor of Germany from 1909 to 1917.

German War Volunteers

On the basis of § 98 of the Army and Military Regulations, every person who has not yet fulfilled their compulsory service can freely choose a troop unit (replacement battalion, etc.) at the start of mobilization. If he fails to do so, he will be commanded during the soon upcoming draft. People who do not have a legal obligation to serve can register as war volunteers with a replacement troop unit, as well as young people between the ages of seventeen and twenty, unless they are in districts where the Landsturm is deployed.

August 1, 1914

2

The student Adolf Reisiger, born on April 1, 1893, in Henthen, was examined today for his fitness for military service.
Findings: Height 1.72
Chest circumference 78/87
Error: H B 85
1A 55 left
1 A 75 flat feet
H = 1
S = with −66/6

Fit for Service

Dr. Jakowski, August 16, 1914
Royal Prussian Field Artillery Regiment 96/
Replacement Unit

3

Declaration of the University Professors of the German Reich

We teachers in the universities and colleges of Germany serve science and do the work of peace. But it fills us with indignation that the enemies of Germany, above all England, should wish to draw a contrast in our favor between the spirit of German science and what they call Prussian militarism. There is no other spirit in the German army than there is in the German people, for the two are one, and we are a part of it. Our army also cultivates science and owes its achievements in no small measure to it. Service in the army also prepares our youth for all works of peace, including science. For war educates them to self-determined loyalty to duty and gives them the self-confidence and sense of honor of the truly free man who willingly subordinates himself to the whole. This spirit lives not only in Prussia but is the same in all the states of the German Empire. It is the same spirit in war and in peace. Now our army is fighting for the freedom of Germany, and thus for all the goods of peace and morality, not only in Germany. We believe that the salvation of European culture depends on the victory for which German "militarism," the discipline of our men, the loyalty and sacrifice of the united and free German people will fight.

Berlin, October 16, 1914

4

To war volunteer Adolf Reisiger
F.A.R. 96 Regimental Staff

Field Mail

October 23, 1914

My dear boy!

You have now been away from us for one night and one day. And by the time these lines reach you, you will probably have long since been in front of the enemy. We don't know how long the war will last, and I would've preferred it if I didn't have to let you go, but on the other hand, I have to understand that there's no rest for you young people, and I can only wish and hope that you return healthy.

Yesterday's farewell weighed heavily on me. Of course, it's all different when, like yesterday at the train station, you are comforted by the thought that I'm not the only mother who has to send her child to the enemy now. And then I had the impression that you were very happy sitting on the train with your comrades. After the train left, we went straight home because Father still had work to do. But I almost couldn't sleep at all. As you know, the noise coming from the station can be heard very loudly from our house at night, and it was particularly bad last night. Transport train after transport train rolled westward. It's impossible to imagine where this great mass of soldiers that Germany is now deploying is coming from and how this whole operation is working.

Last night in the newspaper, there were reports of particularly fierce fighting northwest and west of Lille. You will understand that I am very worried about whether your transport will end up being unloaded there.

I only spoke to Father briefly this morning, but he asked me to tell you that he is very proud to know that his boy is now also fighting (I would have preferred if you could continue your studies). I spoke to the mayor this afternoon. Everyone thinks that the war will for sure be over by Christmas. Please write as soon as you receive these lines and don't forget that you promised to send me a sign of life every day.

With love, your mother

5

The letter was handed out to the war volunteer Reisiger when he reached the headquarters of the active Field Artillery Regiment 96 on a replacement transport.

They arrived in the afternoon. Dog-tired, the commando was herded into a garden and lined up in two rows, facing a large white villa, while the officers sat on the veranda drinking coffee. *Where is this war?* thought Reisiger. *Are we at the front?* Two days ago, we had to leave our railroad car. No train was allowed to continue. And then walking through ruined villages. And then, at night in the barn, there was a dull trembling in the ears: Can you hear the shooting? And now? The officers in their field-gray coats without weapons? And where is the artillery? Where is the enemy?

"Attention! Eyes right!"

An older officer comes down the steps of the veranda, and younger officers follow him. The regimental commander. He is carrying a thick notebook in front of his chest, and behind him are the battery sergeants. *Now they will greet us*, thinks Reisiger. As reinforcements, as comrades who come to the aid of those fighting.

No, the commander orders the men to be at ease. He lights a cigarette and musters the new arrivals. But he does not say anything. No word. He finally waves his hand. "Proceed, Batteriewachtmeister!"[2]

What is coming now is like an auction of unnecessary and annoying goods. The sergeants walk the front row, tap one and the other on the chest, and say: "You, to the First Battery." "You, to the Fourth." "You, to the light column."

It goes back and forth; they are uninterested and unfriendly.

Reisiger watches as all war volunteers get assigned. One after the other they leave his row and step to the side. Only he is still standing. Ends up standing all alone. *Did they forget me? I volunteered here, after all. It's not possible for all the sergeants to just walk past me. The other soldiers are already leaving . . .*

The fattest sergeant sticks his fat index finger first into his collar and then into his belt: "Volunteer, isn't it? I see." Reisiger gets red. *Am I a museum piece? Everybody grins at me.* He looks at lots of broadly laughing faces. He has to swallow to hide his excitement.

The fat constable gives him a shove in the chest: "Fine, I take you with me. Maybe we can still turn you into a soldier. Light ammunition column No. 2, understood?"

With that, the sergeant turns away and talks to the lieutenant who is standing nearby.

For the first time since Reisiger became a soldier, he feels like he is completely alone. And far too young and completely helpless. So, that is what a soldier's life is like at the front? What is camaraderie in the face of the enemy?

---

[2] Batteriewachtmeister (*Battery-Constable*). Most senior NCO rank in a field artillery battery equivalent to master-sergeant or sergeant major. Responsible for day-to-day organizational matters in a military unit.

He still stands at attention and stares at the white villa. The officers return to the veranda, bored. The fat sergeant follows them. After a while, a soldier with a beard arrives. "Well, come on comrade," he says. "You carry the rifle, and I carry your knapsack. I go ahead."

6

Didn't everything turn out the way it was expected? The good, noble, loyal German; the black, vile Russian who wrongly bears the honorary title of a European; the Englishman, suspiciously waiting, and in the south the Balkans, throwing bombs, killing, and betraying—all as predicted! One can regret it politically—but don't you have to bless a people for allowing themselves to be deceived out of loyalty and trust? Today, in our mechanized times as well as centuries ago? There is no enthusiasm for war, no fire as the Latin people show, and no sign of momentum of ever-elated souls. But there is a sign of male defensiveness: simple, noble, silent, and almost bold. It seems to me that the motivation that drove these rigid and inert German people into such an unseen motion was, to the highest degree, moral.

(Emil Ludwig "Der moralische Gewinn" *Berliner Tageblatt*, August 5, 1914)

7

Position of the 2nd L.M.K.: a small village South of Arras. Reisiger and his comrade step out of the garden of the staff quarters and into the street. Here, there is no war. Children and women walk by; they sit in front of their houses, laugh at the two soldiers, and greet them with the German words "Gutt Evenink."

"You will like it here, we have a decent life," the comrade says to Reisiger. "Well, you will be tired; get a good night's sleep first." Then he starts talking. That he has been out here since the beginning of the war. He talks about his family. He is a beer coachman from the Harz mountains. His name is Franz Zeitler. Before the war, he had two horses from the brewery, beautiful animals. To leave the horses was harder than to leave his wife and the five small children. "The old woman bitched all day . . . here, we have peace and quiet. The war has its good sides."

He pushes Reisiger into a doorway. "This is our apartment. It used to be the school."

A whitewashed large room with a blackboard on the wall. The school desks are missing. There is a table in the middle of the room, some chairs, and large boxes around it. Behind it is the catheter on a high platform. Two soldiers are in the room.

Reisiger closed the door and said, "Good evening." Nobody answered, and the two soldiers did not even look up. They had field cups in front of them, sausages wrapped in paper, and were eating their supper. Each had a knife to cut the sausages and bread into small cubes that were pushed into the mouth. Reisiger sensed a helpless weariness. He also felt a red heat in his face. Embarrassed, what should I do now? Say good evening again? Or introduce myself? Or just shake everyone's hand?

Then the mute society suddenly got moving. Zeitler had unwrapped a large, fat piece of pork from a newspaper. That was shocking. The silent men woke up from their slumber.

"Franz, you're really living the fat life again," said one of them.

"Franz has a new bride, you can tell," said the other.

Zeitler puffed himself up and stroked his beard: "Well." He cut off a long strip of fat and pulled it down his throat with bare lips. He swallowed and licked his fingers. He noticed that Reisiger still made no move to unpack his supper. "How, aren't you hungry?" he asked. And then: "Man, we forgot to pick up food for you at the regiment!" He stood up. "Stay calm, Dad has something for you."

He got a second box and pulled out a sausage. "Here, now dig in, we're not among poor people here. There is bread. Eat."

Reisiger thawed at this kindness. He ate. He ate without looking up. It had not tasted this good for a long time. Meanwhile, the others had carefully wrapped the remains of their food in newspapers. They lit cigars and held their heads in their hands. And then an interrogation began.

"You're a student, aren't you? Well, you won't get much from our sergeant. He hates students . . . I am a milk trader."

The person who said that was called Julius Stöckel. He looked like a walrus. He had small black bristles on his head, a long, hanging black mustache, and mischievous little eyes that looked around merrily. He seemed to be the wit of the place. As the conversation progressed, he became more and more animated and finally told his marriage story in great detail. At points that seemed particularly funny to him, he hit Reisiger's thigh with a bang or passionately scratched his own head with the open pocketknife with which he had previously eaten.

His main companion was Robert Strümpel, a baker with water-colored eyes, pale, and a bloated face. He behaved nobly.

With every word, he emphasized the difference in status between himself and an ordinary milk trader. His Hannover dialect helped him with it. The most important thing he had to tell Reisiger was the story of his war marriage.

Touching. If you believed him even 50 percent, you at least had the right to imagine that this Meister-baker had taken his wife from a ruling royal house. And now this delicate girl was running around day and night in silk dresses despite wartime and despite the supervising of the bakery.

And Zeitler? As he listened impatiently to the end of Strümpel's vivid story, he felt animated. In his house they speak another language. He has a quarrelsome wife. A big beast. You had to spank her once a day; otherwise, you couldn't get along with her. A sister-in-law lives with them. A sister-in-law as beautiful as the sun. Good and so on. Only when the candle on the table threatened to drown in the lake of wax did he get up. Time to go to sleep.

For Reisiger, a new perplexity arose: How do you go to sleep here in the field? In the garrison, they had learned that on the frontline, a soldier must at least keep his uniform and boots on. And they must not take off their belt. And now? He observed the others.

They did not think about following the rules. Before dinner, they had already taken off their uniforms. Now, they took off the boots and hung them carefully on nails that were driven into the wall at the headboard. Then they spread out a saddle blanket on the straw, and when everyone had stretched out on it, panting loudly, they wrapped themselves with a second piece.

Reisiger only had one blanket from the garrison. Should he lie down on the bare straw? But Zeitler had some advice. "You, Reisiger, tomorrow I'll steal a nice saddle blanket for you. Today we'll both just lie under my cover. Come on, go to bed."

Reisiger took off his boots. Out of them for the first time in five days. His feet burned. But he felt comfortable. It was just a little unusual with a stranger under the same blanket. He fell asleep quickly, anyway.

In the night, he woke up. He heard a dull rumbling in the distance. Every now and then, a slightly brighter noise. He was awake immediately. He would have liked to sit up, but he didn't want to disturb Zeitler. *This is the war, and that must be the front*, he thought. An intense longing came over him. Forward, forward! When, after a few minutes, the dull rumbling had not ceased, he overcame all shyness.

He quietly pushed the shared blanket aside and felt his way to the window. The window cross stood black against the sky. There was nothing to see outside. Finally, he climbed onto the windowsill. From time to time, the horizon flashed with a reddish light; sometimes it was illuminated by broad white stripes. Reisiger stood up there for a long time. Only when the cold shook him did he crawl back into the straw.

8

Our crown prince telegraphs for rum!

12 cups of good tea with rum packed in a field post letter,
15 of this kind of field post letters for resale,
1 postal package: 9 Reichsmark
Thuringia. Ess. Fabrik, Berlin

(*Berliner Tageblatt*, September 30, 1914)

9

The next morning the service begins. And rolls day after day according to the same pattern. Roll call is held soon after the morning coffee. It usually consists of waiting ten minutes until the master sergeant appears. Activities are then arranged for the morning, especially the usual washing of the ammunition wagon. This wagon is never moved and, therefore, has no chance of getting dirty, but that doesn't matter.

From 10:00 a.m. to 1:00 p.m., it is doused with buckets of water and carefully polished with cleaning cloths that have to be boiled the evening before. And when the gray paint is finally rubbed off in one place due to the extensive use of water, it is ceremoniously replaced with new gray paint. This gray paint is again treated with more buckets of water every day from 10:00 a.m. to 1:00 p.m. for eight days, rubbed off, and replaced with a new coat. This is called military service.

Reisiger became more depressed every day. Soldier? War volunteered?

10

The Reich Chancellor opened the session with a short speech. He welcomed the commission and described the war situation on both fronts as quite favorable. He only wanted to make this short statement today, as he wanted to make a more detailed statement in plenary tomorrow. Of course, there is still a lot to do. He hoped that the Reichstag would once again show full unanimity, since it was precisely this unanimity that would be most suitable for encouraging the troops to make further, highest efforts. The Reich Chancellor appeared quite confident. Of course, he pointed out the possibility of a longer war and advised the German people to tighten their belts when the time came. But at the same time, he expressed his firm belief in the eventual victory. His urgent wish was that the unity of the Reichstag could once again shine into the world.

(*Vossische Zeitung*[3] No. 611, December 1, 1914)

---

[3] *Vossische Zeitung*. Berlin's oldest newspaper with national recognition, representing the positions of the liberal bourgeoisie. The owners Ullstein & Co decided to cease publication in March 1934 to avoid Nazification under the *Schriftleitergesetz* (editor law) of January 1, 1934.

11

War Ministry, December 2, 1914, No. 4141/14 G.K.M.

The reproduction of the Reich Chancellor's statement in Vossische Zeitung No. 611 on December 1, 1914, regarding the duration and consequences of the war is distorted and is based on a gross breach of trust. Suppress further printing and discussions. Confiscate copy of the newspaper No. 611.

12

The foreign policy directed by the Reich Chancellor on behalf of His Majesty the Emperor must not be disturbed or hindered by any open or hidden criticism in this critical time, which will decide over a century. Expressing doubts about its solidity damages the reputation of the Fatherland. Trust in it must be raised and must not be shaken, just like the trust in the military leadership.

(Compilation of censorship orders from the War Ministry, the Deputy General Staff, and the Chief Censorship Office of the War Press Office. Guidelines, No. 3620/14 g.A.1)

13

Against the Naysayers

The deputy commanding general of the Seventh Army Corps, von Gayl, publishes in the newspapers of his district a warning to persevere and trust. "Is it true, he asks, that trust begins to waver here and there? That doomsayers are at work to make people feel queasy and dampen their cheerful confidence?" And he gives the following answer: "If that is so, then it can be said with all clarity: neither now nor ever do we have any reason to allow ourselves to be deterred in our confidence in the happy outcome of the war. Forty-four years ago,[4] our

---

[4] The Franco-Prussian War of 1870–1871 leading to the founding of the German Reich under Chancellor Bismarck.

sword did not rest for seven months. But today, the conditions of warfare, the number of fighters, the expansion of the fronts have grown immeasurably. Enemies all around! But the reckoning with them, in which loyal allies are helping us, is truly well underway. We conquered Belgium by storm, following the commandment of self-defense. Our troops stand invincible on foreign soil to the east and west, our ships are the terror of the enemy. A war in which every day brought a new victory, in which there was no change, no setback, would be a strange war indeed. The best guarantee for a happy outcome is the wonderful spirit of our troops. The closer to the enemy, the fiercer their courage to fight, their enthusiasm, their will to win. And among us, who live behind the front as in the shadow of peace, should anyone despair? Let everyone do their duty to an increased extent and, above all, help economically to strengthen our military armament. Then we can all enter the new year with firm confidence in the victory of our good cause. God save the Emperor and the Reich!"

General von Gayl is quite right in calling for firmness. But perhaps this sermon against the naysayers would not have been necessary if individual circles had exercised a little greater restraint in distributing advance praise at the beginning of the war.

14

Spreading Untrue News of Victory Is a Punishable Offense

The General Command of the Tenth Army Corps informs the Hannoveraner Kurier: Various recent events make it necessary to expressly point out that the distribution and dissemination of unverifiable, false news of victory also falls under the penal provisions of the announcement of November 15, 1914. They are highly capable of unsettling the population and shaking confidence in the Supreme Army Command. The authors of such false news will be dealt with unrelentingly; unless the law stipulates a higher prison sentence, they will be punished with prison for up to one year. The imposition of a fine is excluded. In several cases, criminal proceedings have already been initiated.

(*Berliner Tageblatt*, December 29, 1914)

15

Restaurant "Central Hotel"
New Year's Eve Celebration

The banquet begins at 9 p.m.
Menu:
Oyster pies
Clear turtle soup
Sirloin cuts with various vegetables
Cold Dutch lobster with Tyrolean sauce
Young turkey stuffed with chestnuts
Escaro salad and steamed fruit
Berlin pancakes, cheese, sweets
Surprises

(Advertisement, December 31, 1914)

16

Circus Albert Schumann Frankfurt
Low Prices

"East and West"
Great patriotic showpiece from the present in four acts:

Act I: The Russian in Galicia
Act II: Germans in Belgium
Act III: Our Heroes in France (war episodes)
Act IV: Attack on a Fortress
The phenomenal final apotheosis
400 actors. Two orchestras. Chorus

(*Berliner Tageblatt*, December 27, 1914)

Chapter II

1

On the night of January 20, 1915, war volunteer Adolf Reisiger was ordered to report to the firing position of the First Battery of the Field Artillery Regiment 96 at 5:30 a.m. the next morning.

Things packed. Lace-up boots in the knapsack. Cookware cleaned. Drinking cup washed. Blanket rolled up. Reisiger set off while the others were asleep. To say goodbye was against military decency. There were no signposts. The guard at the end of the village also had no idea about the firing position 1/96. He gave the advice to always head to where the flares rose and where, at times, a reddish flash was followed by a dull thud.

Reisiger was full of excitement. He knew that comrades were in front of him. But next to him stretches only black uncertainty. What is the "front?" What is "the enemy" that is lurking somewhere? Near or far? And whose tentacles cannot be estimated.

The wide road, with poplars to the right and left, was a reliable guide for the time being. Reisiger marched on. At one point, he was overcome with fear. A noise broke out behind him, approaching at an absurd speed. Reisiger looked around, saw nothing, listened harder, and jumped behind a tree to the left. An unlit motorcyclist sped past him. Peace and quiet followed. An hour's march had turned into two.

The clock showed 5:10 a.m. The white flares had long since given up competing with the black sky. The dull thuds had also stopped. It was absolute silence. Nothing moved. The feeling "I'm fine" took over in Reisiger. He walked faster. He would have liked to sing, but a paragraph from the service regulations appeared in front of him that said that one was not allowed to sing, not speak, or even smoke in front of the enemy.

There was a short, hard bang near his ear. A singing sound passed him, almost like a bird chirping. The sound ended with a hard hit on a tree over there. That was a rifle shot that succinctly shattered all thoughts of well-being or smoking. Reisiger put his hands in his pockets and walked even faster. Finally, he saw a dull, yellowish light close to the chaussée on the left. After a few minutes, a half-loud "Halt, who's there?" He had reached 1/96.

2

Captain Mosel was terribly bored. Boredom has been his main activity for four months. The cursed trench warfare. What is an activity that is appropriate for decent active officers? If you still think about September 1914, there, on horseback, of course, he went out every day with hurray, ordered to unlimber[5] and fire twenty, thirty shots in direct aiming. Oh, how the English had jumped out of the grain heaps with their long legs that suit this breed so well. And of course, they had been turned into pulp. And limber again, chasing down the hill, galloping to the next, and then the same spectacle three or four times a day. Now, here we were stuck. Even high patent leather boots and the gray uniform with the black lapels were no consolation. The enemy hadn't fired a shot for weeks. Even worse, you weren't allowed to shoot for weeks either. They lay three kilometers opposite each other as if they were friends and, by friendly agreement, had fallen into a blissful hibernation on both sides. Pointless!

Now, there were orders to perform exercises in the firing position and to carry out gun drills. Mosel called this sport "war in underpants."

At night, it became more bearable. The battery set up an observation post in the trench; they dug in from 11:00 p.m. to 4:00 a.m., and then almost all the squads were on their feet. Well, that was movement; it just had a breath of life in the field.

Mosel crawled around among his men every night until four o'clock in the morning, appeared at the entrenching squads, snorted orders between his teeth, disappeared, appeared with the infantry in the first

---

[5] A limber is an artillery piece on a two-wheeled cart. To unlimber (German: "abprotzen") is the process of uncoupling the gun from the horse-drawn wagon (German: the "Protzen") and bringing it into firing position. To limber means to attach the gun to the Protzen in order to move position. A Protzen has six horses with one rider on each pair of horses.

line, and was constantly here and there. He resembled a hunting dog, nose on tracks, where is the prey, where is the victim, search, search!

3

1 F.A.R. 96 consists of six guns. They have been involved since the beginning of the war and are still all unharmed. You can see the impact of shrapnel bullets on their protective shields like pockmarks. But the sheet steel was reliable, and at Le Cateau it was confirmed—the enemy's ammunition was generally considered useless.

The guns stand about twenty steps apart and each has an ammunition cart to its right, each next to the other, almost in a row. They are not dug in but stand on bare soil. Captain Mosel doesn't think it is necessary to have entrenching work done. The feeling of cover would lead to cowardice.

In his opinion, earthworks are very easy for enemy aircraft to detect. The only thing that happens is that you throw some sand against the lower shields and that you spread large tent panels over each gun and each ammunition wagon as cover. And now, if you saw it from afar, you could believe that you have gray-brown mountains of a lunar landscape in front of you.

One day, the regimental commander visited the position and expressed doubts as to whether the battery, as a brown and therefore conspicuous stripe in a large sugar beet field, could not be recognized by airmen. But nothing will be changed anymore. A few beets are beheaded, and the heads, stems, and leaves are carefully transplanted onto the tent panels.

The troops have their quarters to the right of the battery. There is hole after hole, the walls made of thick clay; the roof made of beet heads on tar paper. Without windows or doors. The entrance is closed by a tent panel. Enough to keep the heat from the small ovens inside. There are no chairs or tables, and the ground, covered with wooden slats only in a few better-furnished rooms, forms a narrow bed. At the other end is the shelter of the battery commander. Mosel hates any kind of comfort. He even disliked having a stove until his feet went on strike. Now, the fire burns, allowing him to write letters on a margarine box.

For three months, he has been reporting to the division each Sunday. He writes that the condition of the men remains good, the ammunition inventory is the same and there is no further news at the front.

4

Reisiger had been assigned to the battery's Third Gun. War volunteer Hermann and reservist Süßkind lived in the same hole with him. This Sunday, they slept until noon, then their stomachs growled. Reisiger, the youngest, had to cook. A tin in the corner of the dugout was full of rice and noodles and soil. He put everything in water and on the stove. When it was cooked, it was divided into three parts and eaten with jam. The recipe was Reisiger's invention. Since even the "old man" Fritz Süßkind liked it, they quickly became friends. This led to a conversation.

Süßkind: "Have you been out here for long?"

Reisiger: "I was in the reserve … Now, I'm in the firing position for the first time."

Herrmann: "How do you like it with us, mein Herr?"

Süßkind: "You can drop the 'Herr' with him. He is no less than us."

Reisiger: "I imagined the war differently."

Süßkind: "Yeah, shit. Man, this is all crap, whatever you're imagining. And I'm telling you, if we don't break through here soon, we won't be home next Christmas either."

Reisiger: "Have you been here since the beginning?"

Süßkind: "Sure, I went out with the old one. He is a fine man, Mosel. He is just so grumpy because there is nothing happening."

Herrmann: "I have been with the battery since Christmas and did not see a single shot. And we didn't shoot either yet."

Süßkind: "And what good is the observation point further out? They only do that so that we stay in motion."

Reisiger: "You don't think the enemy will break through here? The newspaper says that something is expected in spring."

Süßkind: "Let him break through. He has no luck with us. My friend, you don't know Mosel very well. Once he's on edge, there's not a dry eye in the house."

Reisiger: "I would like to see that."

Süßkind: "Just don't push. You can't have it any better than here, I always say. What more could you want? We have our 'apartment,' we have our food every day, and we have no worries. Believe me, my friend. A whole bunch of us had to sleep in cigar boxes at home. And here, well, they can really stretch out. And, after all, a heroic death is one of those things. Back then, in September '14, when we had the last battle, we lost around forty men. There was an old driver who had his head ripped off. And he always said that he'd rather be a coward for five minutes than dead for life."

Everyone laughs. Adolf Reisiger thinks of the many articles by war correspondents in which they always talk about the indestructible humor of our brave, field-gray soldiers. Voila!

The conversation does not come back to life. In the meantime, Süßkind has rolled onto his side and is asleep. Hermann, the war volunteer, takes a sheet of paper from his cap and begins to write a letter. If you walked from dugout to dugout, you could see that almost every man has his cap on his knees and a piece of paper on his cap and is writing with a more or less practiced hand.

5

Letter of Captain Mosel to His Wife

My dear Leni! It's Sunday afternoon and dead boring. A fortnight ago, Leutnant Keller was detached. Or did I already tell you that? Now I'm sitting here in knee-deep dirt in a stupidly boring hovel and can't speak a decent word to anyone. I'm the only officer, and I'll quickly submit a request that a fellow sufferer be sent up here to my position. First Leutnant Busse, who also belongs to our 1/96, changes position with me every three days, and he's now camping down there with the carriages. What do I do? I have a bottle of sparkling wine sent to me every morning with the food truck and drink to your health. Other than this, I envy your brother Karl. In the East, it is a different spirit. I can already see that they're going to shut up store over there. And we're standing still here. It's still very much a question of whether I'll be promoted to major as planned. No sign of the Iron Cross either.

Letter of Vize-Wachtmeister Michaelis to His Bride

My dear bride, I received your lovely field post parcel with cigars yesterday and thank you very much for it. The cigars are good. We live bon and in luxury, like a god in France. I wish you could live like this at home as well. Plenty to eat and no work to do. If you asked me, the war could continue for another ten years like this. I would certainly be promoted to Feldwebel Leutnant[6] and we could marry. Do not send more woolen underwear. I still have two pieces from you and three more that came at Christmas with the field gifting transport. But we don't have a very cold winter.

Notes of Reisiger

Firing position at Arras. I am very happy. Nice comrades. Yes, there is really such a thing, like a big family, also here at the front. The captain is strange. He does not pay attention to me at all—he did not even interrogate me when I had to report to him. He seems to be very popular. Now it's afternoon. What do they do in Germany today? I will sleep a bit.

Report of Field Artillery Regiment 96 to Division

1/96 observed and reported entrenching work by the enemy at point 308. Otherwise, nothing new in the section.

Report of Division to A.O.K.

No news in the section of the Division. Last night, the enemy tried to drive a sapper in front of the 6. Company I.R. 186. The company opened fire. The enemy attempt must be considered as failed.

---

[6] Feldwebel Leutnant. NCO rank for non-active or reserve soldiers. In the event of war or mobilization, the "sergeant lieutenants" were to be used for administrative service or as squad leaders.

Army Supreme Headquarter, January 20, 1915

At Notre Dame de Lorette, northwest of Arras, a 200-meter-long trench was conquered from the enemy. Two machine guns and several prisoners were taken.

6

The Sunday feeling ended with sundown. The groups come together for night work. This time the men get more help, since the construction of the observation point in the forward trench has to be completed within forty-eight hours! The battery is left behind with only two men per gun, with the addition of Wachtmeister Conrad and two sergeants. The others silently set off along the train embankment. Soon, everybody has to carry railroad sleepers, rails, and barbed wire.

Reisiger is excited. After the calm Sunday, he feels a sentiment of joy to be a soldier here at the front. And the feeling turns into pride while trotting silently behind his comrade. *How poor is life in the back with the support units*, he thinks. But there have to be men there too, and yes, they also do their duty, but when you are a soldier, you have to be in front of the enemy!

He is carrying a rail together with Kerner. They let it slip down to the ground when they arrive at the observation post. It is a muffled rumble and the only noise in the silence. The night is starry. You can follow the white chalk line of the trench for a long time in both directions.

So, this is the first trench? Reisiger is curious. *Where is the infantry?* Nothing to be seen. Maybe some vague shadows move in the moonlight. That's all. And where is the enemy?

Reisiger reaches up with his hands and pulls himself up on the wall. He finally sticks his head carefully over the embankment. Again, a white line, narrower and darkened in places.

"They'll shoot you in the face if you don't duck your head again," Kerner says, who stands on the bottom of the trench and yawns.

"But I have never seen an enemy trench."

"Speak quietly, you idiot, or do you think nobody over there has ears? You can hear from miles away at night. At Christmas, we even heard the bells from Arras. Be quiet for a moment and listen."

Kerner stands next to Reisiger, and both listen. Reisiger holds his breath and feels that his heart is beating against the limestone. Strange. Yes, you can hear noises in the distance. A dog howls hoarsely a few times. At one point, the whistle of a locomotive tears through the air. There is also the jingling and rattling of wagons. So, that's the enemy!?

But not for a moment does the consciousness of the "enemy" arise in Reisiger. It all sounds like a peaceful life. Barking dogs, a locomotive whistle, noises from wagons. This is almost a vision of home. In the summer at home, during his holidays, when it was warm and you couldn't sleep because of the strange excitement of an August night, then there were these noises too. Or even the confused dreams of childhood encompassed everything. The neighbor's dog was disturbed by strange footsteps; the train started its journey into an unknown country; the carriages drove shaking through the night; the old carriage in which the doctor was often called to the sick, a wagon with tired people coming from the summer dance, or hay wagons that had missed the sunset and were longing for the barn. Reisiger burns his eyes into the distance.

"You, Reisiger—"

"Let me stay here."

Kerner pokes him in the side. "Don't be foolish, man! We have to go back and carry new planks. If the old one catches us, he'll scream at us. Come on."

He jumps down into the trench and pulls Reisiger by his uniform with him. "Now, let's get on with it. You know, each group has to carry five times. If we don't hurry, we'll still be here tomorrow morning."

Off they go. When they dropped their loads at the observation post for the fourth time, Reisiger gasped: "It's damned hard work." He feels as if both shoulders are slumped and sore. He leans exhausted against the wall of the trench and chews on a blade of grass.

Kerner wipes his forehead with the back of his hand. "You can't do anymore? Yes, boy, what you learn in the cradle..." He turns away. "Well, come on. Or do you want to stay here a little longer? I'm going now. You can join me later."

Reisiger is all alone now.

"Yikes, this drudgery. And thirst, I wonder where the infantry is. There must be something to drink." Reisiger does not think for long. *What could possibly happen to me? After all, I'm a young soldier and can get lost.* He takes a few steps to the right, into the winding trench. Then he stops, startled. In front of him, there's a person. He doesn't move, doesn't even turn his head toward him as he clears his throat quietly. *I see an infantryman!* Reisiger says good evening shyly. The infantryman turns his head a little to one side. "You must be from the artillery?" That's where they get talking.

"So, so, an artilleryman, war volunteer, first time up here in the trenches? It was a real treat for the old warrior." He begins to talk, to cut open. "Yes, the war up here, a great thing. You artillerymen have no idea how we're doing. Stand in front of the French day and night. My dear, you can experience something."

Reisiger understands very little. As he looks at the man, wrapped thickly in a coat, his rifle leaning against the wall next to him, he cannot understand why the war should be so fierce up ahead. The first enemy trench is only two hundred meters from here. Strange, that's really no distance at all.

And danger? He has the feeling of complete safety. Two hundred meters?

"You have to be on your guard here; otherwise, the Frenchman will suddenly come over and grab us," says the infantryman. "Of course, we also have heavy losses," he adds.

Losses?

Reisiger listens. No shot. *Casualties and no shots? Oh, the guy's fibbing. Besides, he speaks so disparagingly of the artillery.* Unpleasant. Reisiger turns around and says goodbye. "Well, it won't be that bad."

When he returns to the construction site of the observation post, no one is there. He sees the piled-up material. The others seem to have left without him. He is gripped by unease. *This causes trouble. Or I must try to catch up with my comrades before they reach the firing position. Should I run across the field? Then I'm sure I'll make it.* He hesitates for a moment. But he quickly puts aside the concern that arises and pushes himself up the back wall of the trench, standing at the top in an open field. He looks briefly in the direction of the enemy. Then he turns his back and walks toward the battery. The path is much less difficult than through the trench. A meadow with short grass and frozen ground. You can move forward quickly.

A few times there are dark-edged craters filled with white pieces of limestone. He looks at the holes in the ground: well, that's where it hit.

He walks for about ten minutes. He stops. He suddenly feels hot. There are people in front of him! He throws himself flat on the ground. He closes his eyes for a moment, opens them again, he checks. Yes, there are people in front of him. He counts. He can recognize three. They lie in a row, two flat on their stomachs, the third kneeling. They have the rifle at the ready; the barrels aimed roughly at him.

He is gripped by fear. Enemy? It can't be the enemy. On the other hand: What should Germans be doing here in the middle of an open field? He straightens up a little and wants to call out to them, but the darkness is broken by a flare. And he sticks his face in the grass. Isn't there an approach trench somewhere? He sees nothing. So, there's nothing left to do but make yourself noticed by the people ahead of him.

"Hello," he whispers. "Comrades, I am one of us." They stay motionless; they still point their rifles at him. He is repeating the sentence, this time a little louder. No answer.

He thinks it is best to crawl to them. That's a maximum of thirty meters. He doesn't feel well about it. What's the use of it? He can't stay here. So, he crawls forward. Slowly. He delays every movement of his body. The soldiers are now no more than six steps away from him. "Comrade," he whispers. No answer. Now Reisiger is getting scared. Some kind of communication must be possible! There are two options: Jump up now and run past these guys toward the battery firing position or jump and lie down next to the first man so that he cannot move his rifle around. The second option seems more reasonable.

Reisiger pushes himself off the ground with his hands and, in the next second, is lying next to the soldier. He moves his head close to his ear and says: "Comrade, you can't sleep here. Imagine if an officer showed up now." No answer. The comrade is still stubbornly holding the rifle in front of him and doesn't even turn his face. Reisiger nudges him. "You, say something, it'll be dawn soon." No answer. Then Reisiger puts his hand on the right hand of the man next to him. It's freezing cold.

Reisiger pulls his hand back and tears himself free. *Oh well, oh well, he's dead.* He looks carefully at the second, the third. They all remain motionless. They have their rifles at the ready. Two are lying on the ground. The third kneels. *They are dead?* No head dropped on the chin. No helmet fell off. Not a finger detached from the rifle. Just as Reisiger had often seen as a child on parade grounds, here lie three infantrymen ready for battle, their faces well aligned against the enemy. Only, alas, they are dead.

The first dead that the war volunteer Adolf Reisiger sees in the field.

He is completely calm. All fear is gone. No excitement at all, no heart pounding. He stays on his knees next to these dead comrades for a few minutes. He marvels at how they look so motionless against the enemy and how they will probably continue to look like this—against the enemy for weeks to come. Then he stands up. He takes a few steps. He turns up the collar of his coat. And then he runs at a wild gallop toward the battery firing position.

He arrives there at the same time as the last men of the entrenchment commando. A head count is taken, and no one is missing. Dismissed! After a few minutes, Reisiger lies in the dugout with the other two comrades and sleeps.

7

In New York, a protest song against the war is now being sung in all the vaudevilles, music halls, on the street, and in the salons, which goes like this:

> *I didn't raise my boy to be a soldier,*
> *I brought him up to be my pride and joy.*
> *Who dares to place a musket on his shoulder,*
> *To shoot some other mother's darling boy?*
> *Let nations arbitrate their future troubles,*
> *It's time to lay the sword and gun away.*
> *There'd be no war today,*
> *If mothers all would say,*
> *"I didn't raise my boy to be a soldier."*

(*Neue Freie Presse*, February 23, 1915)

8

How Long Will the War Last?

Wire Report

Amsterdam, February 4

The Times *reproduces a question from the* New York American *about the duration of the war. The German Ambassador, Count Bernstorff, replies: "If I say that the war will last a long time, it is immediately said throughout the country that I have said that Germany wants war. If I say the war will be short, then it is said that Germany wants peace. So I'd rather not say anything, because it's being distorted either way."*

Richard Bartholdt, member of the US Congress and founder of the Neutrality League, explained: "I believe in Germany's triumph, but as an American and a member of Congress, American neutrality is the most important to me. We claim to preserve it, but our actions speak a different language. The French General Bonnal said: "The war will probably last a long time, because the German is proud and a good soldier. The allies still have a lot to achieve, but they will succeed."

(*Vossische Zeitung*, February 5, 1915)

9

The next day, in the firing position, around midday, the captain is in a very bad mood and has the idea of holding a gun drill himself. He personally shouts the command shortly after lunch is finished: "Ready to fire, the whole battery!"

The dressage begins. Reisiger is assigned as the gunlayer. It's more fun than in the garrison. The captain gives very good commands. The six guns are prepared to fire in no time. Everything works fabulously. Next to Reisiger is Sergeant Gellhorn, the gun commander. He makes sure that Reisiger is operating the targeting mechanism correctly. He is very satisfied. The captain's commands follow each other so quickly that the whole battery is sweating after a few minutes. Then, there is a break in the firing. They rest. Suddenly, you hear a bang in the distance.

Very quiet, very far away. The next moment, Reisiger sees that there is a commotion in the battery. Some of his comrades stretch their necks and stick their noses in the air, sniffing. Others huddle close to their guns. Gellhorn puts his arm around Reisiger's neck. Everything just takes seconds. Then, a piercing, glaring whistle erupts from the air. Shrill. And then a roaring crash. Gellhorn lays his body on top of Reisiger and presses him heavily against the gun shield. Both almost fall over. He hisses through his teeth: "Damn hogs, the enemy is shooting."

Reisiger feels his pulse throbbing in his throat. *So, the enemy shoots? So, damn hogs?*

When he notices that Gellhorn is standing up again, he raises his head. He can just see that the heads over there, by the other guns, are also going up again. He turns around. About two hundred meters behind the battery, a cloud of smoke rises from the earth with an elegant white torso and a black, disgusting head.

"These bastards shoot grenades," says Gellhorn.

Captain Mosel's bad mood is gone. He has a broad smile on his face. He pushes his silk cap far down his neck, puts both hands in his trouser pockets, and steps close to the battery. "Thank God these stupid monkeys are finally showing mercy again." Then he sniffs the air expectantly.

Reisiger looks at the neighboring gun. The faces of the comrades are less cheerful. "Do you think there's more to come, Herr Unteroffizier?"

Gellhorn: "I suppose this is your baptism of fire?"

Reisiger: "Yes, Herr Unteroffizier. I haven't quite grasped it yet."

Gellhorn: "You don't need to be curious about that. You'll learn ... Here comes another carrion."

Everyone makes their bodies small. The noise tears through the air. Again, *bang!*

"That is ahead of us," says the captain triumphantly. "Watch out, boys, they'll have us in a minute."

Reisiger trembles. *So that's it? This is the baptism of fire?*

It's like sitting on a platter. There are no guarantees that the next shot won't actually hit the battery.

Now, Süßkind pushes himself between Gellhorn and Reisiger. "Are you afraid?" he asks. He makes such a pinched face that Reisiger feels sorry for him.

Reisiger fumbles with the targeting mechanism: "I don't understand why we don't shoot back?"

Süßkind spits out: "Then they really get to us, right, Herr Unteroffizier? The captain should let us go."

Reisiger looks at Gellhorn. He also wants to say something. There is a new explosion. Much louder than the two times before. It hails turnips and earth over the gun.

Reisiger, Süßkind, and Gellhorn fall to the ground and don't move for a few seconds. Waiting. Waiting. Finally, Gellhorn carefully looks over the protective shield. He points to a black hole from which slimy, yellowish smoke is flowing. "Damn It! It was only twenty meters away."

There is no time for further discussion. Mosel's voice, now very sharp and piercing, suddenly pulls the whole battery together. "Now it's our turn," he shouts. "Shrapnel fuse, whole battery 3,600!"

All thoughts are gone. Quick movements. Six messages: "Gun ready to fire!" The captain stands on tiptoes, then in a squat. "Battery, fire!"

With a mighty roar, six barrels hurtle backward. The guns rebound with a short jerk. Then a pungent cloud envelops them. Everyone listens. The explosions thud in the distance.

"3,100 fire! Aim, load, fire!"

Five times the same thing. Five times the explosion in the distance. The enemy does not respond with a single shot.

"Battery . . . hold." Mosel takes his cap off his head and walks slowly back and forth, satisfied. "Now they'll probably be quiet. We're going to continue with the artillery drill."

But after a while, he gets bored. "Sergeant Conrad is taking over . . . . Sergeant, let them do the aiming exercises for another twenty minutes, then the battery can leave." He puts his finger to his cap and steps into the shelter.

Sergeant Conrad lifts his hand: "Shrapnel fuse . . . half left . . ."

He does not say more than that. There is an unexpected new whooshing sound in the air. Sharp, loud, louder, louder—impact!

Reisiger is thrown under the ammunition truck. He braces himself and looks up. Conrad is right behind him. He is rolling around, panting, and groaning like a sick animal. He raises his arm and lets it fall.

Reisiger sees: Conrad's left hand has been shaved off at the root. A thick fountain gushes from the stump. There are calls for medics from all sides. The captain is back, kneeling beside Conrad and saying:

"Battery dismissed." The sergeant is placed on a stretcher and taken to his dugout.

Reisiger's knees are shaking, and he feels something choking in his throat. *So, this is the war?* There stands a man, loud and strong, with provocative courage. The sun is shining, and the sky is blue. Suddenly, the man is lying on the ground, and blood is splattering. The man will go home and will never have a left hand again in his life. That is disgusting. The others in the dugout are also depressed. Conrad went out with Süßkind at the beginning of the war. A dashing boy. "It's a shame we're losing him, because you have to remember that you could get a sergeant who will hound us day and night."

Not much more is said until it is dark. Reisiger takes a cigar out of his pocket and gives it to Süßkind. Then he shakes his hand. He notices that this is the first time he has shaken hands with a comrade out here in the field. He takes his bag and reports to the captain. At night, he is back in his quarters in the column. The others are already asleep. He lies down with them.

10

From the next morning onward, Adolf Reisiger only thinks about the fact that he will be back up at the battery in three times twenty-four hours. Over coffee, he talks about his baptism of fire. Then the wagon driver arrives, and Reisiger is ordered to sweep the streets. This is a new job. He is eager to find out how to sweep roads in accordance with regulations.

There are already seven men outside in the battery parking lot. Four of them are holding shovels that have been requisitioned from villagers. He and the others are given long-handled turnip hoes. Under the supervision of Sergeant Gaensicke, they head out onto the village street.

The eight street sweepers are all war volunteers. Even the otherwise fine gunner von Oertzen, an assessor in his civilian life, has to spend the next few hours scraping the slightly thawed and muddy dirt from the road and pushing it into piles, which have to be neatly beveled on all four sides and polished at the top like a sheet of glass.

Reisiger is annoyed. The whole village is teeming with civilians. There must be a hundred older men and several hundred younger women and girls who could keep the place clean. So why soldiers? But orders are orders. Gaensicke is so convinced of the necessity of this mission of war volunteers that he can't help but shout at Reisiger as he inspects the already completed piles of polished dirt. There was a big piece of straw in the second pile from the left that didn't belong there. But war volunteers were no good, even for sweeping dirt. Gaensicke also wanted to make sure that the whole company practiced day after day until the piles finally looked the way the column leader wanted them to.

They arrive at the end of the village street around midday. Gaensicke orders the eight men to line up in two lines, shovel and hoe over their shoulders. He leads them in step through the completed work of which he is proud to be the creator. He counts the individual piles out loud and then reports to the sergeant: "Back from cleaning the streets, 178 piles." Gaensicke makes a face as if he wants the Iron Cross immediately, preferably pinned to his hero's breast by the commanding general himself.

At lunch, Reisiger bangs his fist on the table. He would go to the captain later and demand his immediate transfer to the battery. He was not thinking of returning to the streets.

The comrades calm him down. Gently, gently. For God's sake, he shouldn't do anything nonsensical. Why get so excited? Why plunge into misfortune by force?

Above all, Franz Zeitler talks to him. He is like a father to him and has the philosophy of an old frontline soldier. He says: "Look, Adolf, it's like this: If you report to the front now and they slap you in the face, you'll always have to reproach yourself and tell yourself that you didn't want it any other way. Believe me; we all do everything that is ordered. But we don't push ourselves. Never do anything that is not ordered. That's the highest order of all."

Reisiger remains silent: *But … am I not a* war *volunteer?*

11

Mindful of his high calling to protect the Throne and Fatherland, the soldier must always be eager to fulfill his duties.

(Article of War 1)

12

Every order must be carried out as literally as possible. If the soldier encounters difficulties in doing so, he must not immediately regard the order as impossible to carry out but must consider how he can overcome these difficulties and perhaps reach his goal in another way. The main thing is that the soldier understands the "intention" of the order, i.e., what is actually important, and carries it out. This is called "purposeful" execution of an order.

(Comment on Article of War 11)

13

Reisiger's wish was not fulfilled for the time being: He was not reassigned to the battery.

The routine of column duties: wagon washing, foot drill, street cleaning.

Volunteering for war? Where is the war? Where is the enemy?

The enemy? One morning, the order came to set up a guard for the local commandant's office. There are French prisoners in the church.

The church is a rough brick building. The pews from the nave were burned as firewood during the winter months. Only the altar still stands, between two red brick pillars. A prayer chair stands in front of its steps. Piled straw lies along the long wall of the entire room.

Reisiger and von Oertzen have the first watch. They open the church door. They see the prisoners. Most of them are lying in the straw and take no notice of their neighbor or the two new guards. They don't speak either. Oertzen and Reisiger walk timidly down the center aisle. The sound of their own boots distresses them. After a few steps, they stop. They don't quite know what to do. Oertzen has turned his back on the prisoners, looks along the walls of the church, and says, half admiringly, half pityingly: "It's actually quite a fine room. Look at the nice windows."

Reisiger raises his head. Gray-blue alternates with lemonade red; no one can claim that the windows are beautiful. He nods and remains silent.

After a while, Oertzen says: "I'll go around the perimeter of the church, because you never know if there's an unlocked door somewhere. And then someone might get out." He is gone.

Reisiger feels the temptation to follow him. He walks up to the door, almost on tiptoe. But then he puts the carbine on his shoulder with a flick of his wrist and turns around very loudly. And then he takes firm steps toward the altar.

Finally, he turns to the side. It's not easy. It takes effort. But, by golly, he didn't come here for fun! And these are prisoners and enemies!

They have strangely small feet. And wraparound leggings that don't quite reach the knee. And funny red pants over them. They really don't look like soldiers. How can a grown man walk around in red pants? And the long blue tailcoats aren't exactly masculine either.

One of the Frenchmen isn't wearing a helmet. His face? The moment Reisiger turns his eyes on him, he bends his head low on his knees.

The next one. He looks straight. Reisiger takes half a step sideways and stands in front of him. Now, two pairs of eyes must meet.

But they don't. The Frenchman's eyes look fixedly and motionless through Reisiger as if he was air.

Reisiger experiences the same thing with the next ten or twelve men: They bow their heads when he looks at them. Or they act as if no one is standing in front of them. And one of them slaps his hands over his face with a loud clap and covers his eyes.

Reisiger doesn't know what to do. Maybe the right thing to do would be to shout at the whole gang. He could demand that they all stand up at his command and put their hands on their trouser seams. At worst, he could give them a good whack in the back with the butt of his carbine. Then they would realize how to behave toward a German soldier.

Reisiger considers all this but does nothing. He remembers that he is twenty-one years old and that there are defenseless people sitting in front of him, most of whom he estimates are a good ten years older than him. This shames him. He turns around. There, someone calls out, "Comrade." He looks along the squatting row. No one moves. A second time: "Comrade!" He looks further. He sees a bundle, red and blue, lying on the steps under the altar. The bundle moves, and a hand rises from it with a very delicate movement.

Reisiger goes closer. A French officer. He is wearing a cap with a gold braid. A pale face with a black goatee.

What now? He understands a few words of French, but he will hardly be able to make himself understood. The hand waves again. He steps close. The officer throws back his coat; his tunic is unbuttoned, and there is a large damp bloodstain on his dirty shirt.

The officer's hand points to this stain. It trembles and flutters back and forth.

Help? Reisiger doesn't have the courage to pick up the shirt and check the wound. He mumbles a few German words and then tiptoes through the whole church to the exit and calls out loudly, "Medic!"

Oertzen has heard the call. He just looks at his watch. "The replacement will be here in two minutes. I don't know what to do there either. I'm sure the others can do it much better."

Changing of the guard. Reisiger and Oertzen report that everything is all right in the church except for the wounded man. They unbuckle their belts and sit down at the table in the guardroom. Suddenly, they come alive. Oertzen tells a joke.

After four hours, they have to take up their post again. They walk with loud steps to the altar. No one calls out "Comrade" anymore.

14

The publication of reports on so-called fraternization scenes between friend and foe in the trenches is undesirable.

(Chief Censorship Office No. 38. O.Z. January 22, 1915)

Chapter III

1

Diary entry of the Gunner Adolf Reisiger:

Sunday, May 9, 1915, four a.m.: Worked on a new battery position from yesterday evening, eight p.m. until half three this morning. Entrenching work, new shelters. A great drudgery. Just now, four o'clock, into the straw. But today is Sunday, and I am not on duty. We will sleep until lunch. I am glad that I'm finally a permanent member of the 1st Battery. We're at Mercatel, south of Arras.

2

On Sunday, May 9, at nine o'clock in the morning, First Battery F.A.R. 96 was alerted. The crews, exhausted from building their positions, cursed and swore as they were taken out of the straw.

Alarm? What nonsense! After all, are they supposed to do drill exercises after so many nights?

They provisionally packed a few things into their panniers and duffel bags, put on their helmets, strapped on their harnesses, and headed for the parking lot. Captain Mosel and First Lieutenant Busse were already there. A Lieutenant von Stork, who had been assigned to the battery since the day before yesterday, also made his appearance.

As usual, rumors soon emerged and flew from gun to gun. "Latrine slogans." Nobody believed them. Only one thing was heard: The battery was going to Russia.

Fabulous. The war is pretty much over in Russia anyway! So now is finally a decent time!

But then suddenly, all conversation stopped. The regimental commander appeared with his adjutant. He rode slowly down the battery, chewing his mustache, looking straight ahead. Mosel galloped toward him and reported. The officers stood to one side and bent over a map. The colonel spoke quickly, his right hand flicking through the air several times.

He broke off, shook hands with the battery officers, and rode off at a gallop across the field.

"Battery march!"

The battery slowly rode out of the village.

The weather was fine. The May sun was shining warmly. They opened their coat collars, sat comfortably in their saddles and seats, and smoked.

Russia? Nonsense. Then the baggage would surely have been moved out too. So, of course, goddamn battery drill at the regimental commander's request. Maybe he slept badly last night.

But where to drill? The field here on the right was the usual exercise ground. Why doesn't the captain turn right?

The battery reached the neighboring village, the quarters of the light ammunition column to which Reisiger had belonged for months.

There was a surprise: The column stood on the market square—ready to march. And it joined them as the battery passed by.

Departure for Russia? Drill exercise?

The gunners murmured with the old people. All doubts were set aside. It dawned: We are going into battle.

We drove for two hours. The sun was high. At this time of day, food is generally served. We were hungry. There was no time to make coffee or have breakfast when the alarm went off this morning.

The battery remained on the march.

We went through unknown terrain. The road was rutted and had potholes with splintered stone edges and cracks. Many columns must have passed through here so that the footsteps of the infantry, the stamping of the horses, the grinding of the guns, and the roll of the trucks had left such deep marks.

They drove through villages. The houses crouched, poor and mysterious, on the sides of the disfigured road. The buildings were almost all blind, windowless, with gray gables; the hedges of the front gardens carefully dabbed with green.

And when the battery rattled and clanged at the entrance, curious residents crept up to the gates, old men who didn't lift their eyes, women with embarrassed smiles or open mockery. Sometimes, children approached the marching column and waved.

The military were nowhere to be seen. There were signs on all the houses with the names of German units, but no soldiers were to be seen.

It was getting to one o'clock, two o'clock in the afternoon.

3

May 10, 1915 *Grand Headquarters*

Western Theater of War

Southwest of Lille, the major Franco-English attack expected in response to our successes in Galicia began . . . The enemy—French as well as white and colored Englishmen—led at least four new army corps into battle in addition to the forces that had already been deployed in that line for some time.

4

The First Battery F.A.R. 96 stands on a road leading to a height. This height rises unconventionally and randomly from the terrain, black against the yellowish sky.

The column is waiting behind the battery, infantry in front of it. Motorcyclists are constantly racing past; riders gallop across the pavement.

Gunfire roars from the horizon.

At one point, a higher staff arrives on horseback at a sharp pace. One catches a word: "Vimy." Another hears: "Loretto." Is there anything to be gained from it? Can it be interpreted? The words go through the whole battery at lightning speed. Loretto? Yes, I've heard it before. Thick air!

Now everyone knows.

The drivers and gunners are dismounted and walk slowly back and forth between the guns. They bite nervously on their cigars. Stomachs are rumbling; everyone has been without food since yesterday. Some have a piece of bread in their boot or coat pocket; this is now shared with the men next to them.

People don't like to look ahead. There, where the road bends, where it hugs the darkness of the heights, black clouds occasionally rise from a village with a muffled crash.

The drivers look after the horses. The animals are tired. Occasionally, one drops its head. But that only lasts a moment; then it tosses its mane up again. Close to the battle.

They chew their teeth vigorously. They scratch the ground. Thirsty and hungry. In the dusty ditch of the highway, they can only pluck the poor, bristly grass.

The officers wait and dismount at the head of the battery.

Finally, a rider comes from the front. Around the bend, gallop. And then, gallop, closer. And halts the horse. Hands a white sheet of paper to the captain.

He reads it. "Battery!"

No one misses the speed. Master Sergeant Hollert, another sergeant, and two men from the telephone squad trot up. "Gallop! Follow the captain! To the front."

At the bend. Swallowed up by a black cloud.

It is four o'clock in the afternoon.

5

At six o'clock, the battery is still waiting in the same place. Tired from sun and hunger. The crews are lying down in the road trench. Some are asleep. Most are dozing off. What will happen?

A view of the village: The fountains of smoke are now spraying up more frequently than before. The noise roars harder through the air. At one point, a house is hurled against the sky; brittle beams explode into flames. Impressive! The village is no more than a thousand meters from here. All we need now is for the enemy to shoot here.

The movement in the lined-up columns swells. There's nothing more cursed than waiting!

What time is it? 6:15 p.m.

Voices are heard from the top of the battery. Aha, the infantry is advancing. Man after man, in single file, pressed close to the sides of the road, one with his eyes on the other's neck. Rifles slung around their shoulders. The footsteps scrape heavily over the gravel.

Where to? It's simply impossible to lead the infantrymen through the burning village, especially, when the fire is getting stronger.

In any case, it's better to be an artilleryman. We can ride at a gallop if necessary. It can't catch us so quickly. Poor guys. Trotting in this heat.

They disappear behind the bend in the road.

6:30.

First Lieutenant Busse is restless. He keeps putting his binoculars to his eyes. Where's the message from the captain? "Stork, can you explain this? I'm beginning to find this matter uncomfortable. I don't know how much longer we're going to stand here . . ."

Von Stork remains silent. He had only joined the field eight days ago, was a cadet, and still had to take his exams. Eight days ago, he was still in Lichterfelde. Nice start, straight into the mess!

He also looks through his binoculars. But he's less interested in the news from the captain than the fire in the village.

Will the battery have to go through there later?

Busse: "Will I just let the battery advance? What do you think, Stork?"

Stork: "As Herr Oberleutnant commands. I just don't think the approach route is easy . . . I mean, if we're supposed to go up there."

6:40.

Shouts from the battery: "Officer coming . . . Officer coming."

A horse breaks out of the village between dust and fire. The rider lies low on the horse's neck. Sergeant Hollert. Gallop, gallop, dashing around the bend, gallop. You can already hear the beat of the hooves, hard in unison. Gallop, and closer and closer. The map pocket swings. The arms are not bent, leaving the reins long. Closer and closer. And he halts, close to the officers, so that the horse bends at the joints. "Report to Herr Oberleutnant: Battery is moving into position immediately. Herr Oberleutnant takes command, I'll lead."

Busse gets on his horse. "Battery mount up!"

"Battery march!"

He thrusts his fist twice into the air. The signal continues from gun to gun: "Battery trot!"

The ammunition column remains behind, awaiting its own orders. 6:55.

6

There is no other way to reach the ordered position than to drive through the village.

As the battery, tightly closed in, stands about a hundred steps from the village entrance, Busse lets the gunners come to him. Shell after shell sweeps onto the road. The crash hits the ears, chasing the blood. "We have to get through! The gun crews are acting on their own! So, let's go! The battery gathers again a hundred meters behind the village on the other side. Take care!"

He spurs his horse so that it mounts. Jumps on. The lieutenant and the sergeant follow. Between two shots that are firing down, they are gone.

There is nothing more to think about. The First Gun gets ready. The gunner stands next to the front rider. He waves once more to the rear. As another cloud of smoke sprays up, he shouts, "Gallop, march, march!" The horses start immediately. When they are twenty meters away from the first house, another shot rings out. The front horses reared up and tried to break out to the side. But the driver was expecting it. And when the next shot hits the same spot seconds later, the gun is already in the village street and races on.

This is repeated. Off goes the Second Gun. The Third gets ready. Reisiger and Jaenisch are sitting on the Third, in the gun carriage seat. The gunner, Sergeant Gellhorn, rides past them.

Reisiger is so tense that it becomes impossible for him to stay in his seat. It is unbearable to have his back against the direction of travel. As the horses gallop off, he turns around and stands on the footrest.

Gallop, five, six jumps!

The ground splashes up. The shot is about ten meters from the gun.

Gallop on! The gunners pull their heads against their chests.

Gallop on, already in the middle of torn houses, smoke, and fire.

To suddenly bring the gun to a halt now is completely out of the question.

Completely out of the question, Reisiger thinks. Completely out of the question. So, if the next shot, the next shot, the next—unpredictable—there—there—then we're all screwed—

Reisiger grits his teeth and clings to the handles of the seat as if in convulsion.

A hot storm hisses at him. A pressure hits his chest. A flame leaps up. An insane roar bursts from it.

The horses are thrown out of the barrel, staggering against each other. They race on.

Then Jaenisch says, "We're through."

The last house in the village is behind them. In front of them was the Second Gun. "Halt!"

If only we could keep going! Waiting here stinks.

To the left: a dark forest. Trees are being felled. Again and again, fire jumps out of the trunks as thick as a man. Then they sway unsteadily, heave, splinter, and collapse with a screech. And the echo shrieks everywhere.

The first three guns still stand together.

It whistles and whooshes and whistles and whooshes and whistles.

The horses snort, showing distorted red nostrils.

The men glare at the smoking forest. What to do? Keep your head down? It makes no sense. If you get hit, you get hit.

Why is Busse waiting? It would be better if the gunners were allowed to bring their guns up to the height at their own command.

Aha, the Fourth Gun bounces.

If only we could keep going!

A shot makes a stupid joke; it hits directly, very close, at the feet of Busse's horse. Half a meter in front of it. Ssssss— Everyone raises their heads. That's it for Busse— *Where's the outcry?*

There's a muffled bubble, a belch. Dud, not dead. He got lucky!

Busse's face is white, his forehead and cheeks gleaming.

"The last gun is here!" shouts von Stork.

Casualties? No.

Neither man nor horse has even the slightest scratch.

If only we could keep going!

No casualties? Oh. That's a pointless coincidence. Danger is a pointless coincidence. Rescue is a pointless coincidence. You don't understand why there are no casualties when everyone had to wait in the middle of the fire. You don't understand at all.

Thinking is stopped: "Battery swings left . . . march!"

Through the forest. Up to the heights. Oh, dear God, what a mess! This road! Hole after hole in the dusty grinding sand, so that the guns stagger and only ride on one wheel for stretches.

Come on, gallop, trees are being felled here, with fireworks and a drumbeat. The gunners held fast. Clutched to their seats. Eyes closed, chin on their breast.

Bloody noise. That miserable howling and tearing in the air. It can make you sick because it stinks of sulfur, suffocating.

The horses are almost sliding on their bellies, pulling.

The ridge is reached.

We pass through a birch grove. At its edge, toward the enemy, stands the communications sergeant who had ridden ahead with the battery troop.

"Battery halt! Forward, dismount!"

Down from the guns, unlimber, tear the ammunition baskets from the guns, canvas, baggage, the whole shebang.

Orders for the Protzen: "Take cover at the edge of the village, somewhere. Horses remain harnessed." Hollert takes command: "Off!"

7

Despite enemies all around, Myrrholin soap is and remains the well-known unique skin care and health soap at home and in the field, as it has been for twenty years. Piece 55 Cents. Available everywhere.

(*Würzburger General-Anzeiger*, April 28, 1915)

8

The battery stands in a line, thirty steps from gun to gun. The small birch trees provide good protection against air reconnaissance.

The observation post is in the infantry trench. The telephone line to the firing position has already been laid. Busse connects with the captain. Mosel: "Pretty dark, isn't it? No casualties? Well, that's good, Busse ... It's quiet up here now. The infantry doesn't believe in attacking. I'll be right back to the position. Send a telephonist up here to keep watch at night .... And then one more thing: I forbid all entrenching work in the battery. Do you understand? No, no entrenching is allowed. So, no digging cover holes either. The planes are supposed to be damn busy here. And we mustn't give ourselves away .... Good."

The war volunteer Raschke is ordered into the trench as a telephonist. He says goodbye to his comrades. This is unusual, and he is teased about it. As he shakes Reisiger's hand (they have come out here on the same transport), he says, "Heavens, Doctor, you're solemn."

Raschke has a frozen smile on his face. He turns around, says, "Well," and leaves.

When the captain arrives at the position, it is dark. The gun crews have wrapped themselves in blankets and are huddled together. It is cold. They are also all hungry. But where are they supposed to get something to eat here?

The captain goes from gun to gun. "Does anyone have any bread?"

No. —Something must be done.

Captain Mosel sends a sergeant downhill. "Tell him to look for the Protzen and order Sergeant Hollert to bring up bread and something hot to drink immediately."

That is a solace.

The enemy fire has ceased. No shots have been fired into the rear area for a long time. So Hollert may see. There may still be infantry down below who can provide provisions.

Yes, after a short time two drivers arrive. They bring a sack of bread and two pots of cocoa. The captain personally sees to it that everything is distributed fairly. "Now, boys, eat and try to sleep. No need to keep watch. Telephone stays manned. You've behaved well today. I hope it stays the same tomorrow."

The mood is good. People wrap themselves up more tightly after dinner and move closer together. Some fall asleep straight away. A beautiful night. With a full moon. With a soft wind in the birch trees. Once there is a plane over the position. A Frenchman, as you can hear from the singing of his engine. He must be very high. It sounds friendly, the singing. Otherwise, there is almost complete silence. Even in front, in the direction of the trench. Sometimes a hand grenade yelps. But that makes no impression on anyone.

The captain walks up and down behind the guns with Busse and Stork. They chat and laugh as if they are on maneuvers. Finally, they sit down under a thick tree. The conversation falls silent.

When it gets light around three o'clock in the morning, there is new movement in the battery. The captain calls Reisiger. He takes him aside: "Reisiger, they've just reported that Raschke is dead. Come with me. We're going on observation. Get some cable wire so that we can keep in touch with the battery under all circumstances."

Raschke dead? The doctor Raschke killed? The first war volunteer?

Reisiger has to swallow. Raschke was killed. That's strange. It was such a quiet night. The few hand grenades. A couple of rifle shots. You could sleep peacefully about it. And now someone was killed. Strange.

"Yes, hand grenade. He was killed instantly." Reisiger hears this as he picks up a roll of cable wire from the telephone sergeant. "Too bad, he was a pretty decent guy," says the sergeant. Then he hands Reisiger some insulating tape. "Yeah, that'll be quick. You'll be gone in no time ... All right, Reisiger, let's go. The captain has already trotted off. Don't get caught. I just heard the morning report from the infantry: seventeen dead. Yes, yes."

9

The Faculty of Philosophy at Berlin University has made a worthy addition to its doctoral regulations. It has decided to award such candidates who have passed the doctoral exams, but who have fallen in battle for the Fatherland before completing their doctorate, the dignity of "Doctor of Philosophy and Master of Liberal Arts" in order to keep their memory alive in the annals of university history.

The amended diploma, the version of which, as we understand, originates from Professors U. v. Willamowitz-Moellendorf and Ed. Norden, certifies the fact of the doctorate after the usual introduction with mention of the Emperor and the name of the former rector with the following words: "The Dean of the Faculty of Philosophy has awarded the distinguished, learned and brave . . . . . . (name), who, having passed the philosophical examination with (great, very great) praise and having written a (very) commendable dissertation entitled . . . . to the pleasure of the faculty, earned for himself a fame beyond all praise through his death. The distinctions and honors of a Doctor of Philosophy were conferred in order to consecrate his memory."

(*Vossische Zeitung*, May 12, 1915)

The parents of the war volunteer Raschke were among the first to benefit from the imitable addition to the doctoral regulations of the Faculty of Philosophy at Berlin University.

10

London, May 9 (Reuters bureau report): According to the rescued from the "Lusitania," it was a bright, calm, and sunny afternoon when the ship was torpedoed. Most of the passengers had just had breakfast and were standing on deck looking out for the Irish coast when suddenly, a white streak was seen approaching the ship through the blue water. A terrible crash followed; the whole ship shook and began to turn in the hope of reaching the coast. Then, it was hit by a second torpedo. It tilted sharply to one side and sank in twenty to twenty-five minutes after the first explosion. The boats on the port side could not be lowered because the steamer was leaning. Some sailors saw a submarine for a moment, which quickly dived and did not reappear.

(*Rheinisch-Westfälische Zeitung*, May 10, 1919)

11

Rotterdam, May 10: The "Nieuwe Rotterdamse Courant" receives a report from London: It is now certain that almost 1,500 people have perished from the "Lusitania."

(*Berliner Tageblatt*, May 11, 1915)

12

*Circus-Varieté Schumann
Small prices.
Daily from 8 to 11 a.m.
Circus Varieté performances and
Torpedoing the "Lusitania"
Smoking permitted.*

(*Vossische Zeitung*, May 14, 1915)

13

W. Official. Berlin, May 14. From the report of the submarine that caused the "Lusitania" to sink, the following facts emerge: The boat sighted the steamer, which was not flying a flag, on May 7, 2:20 Central European time afternoon on the South coast of Ireland in clear weather. At 3:10 minutes, it fired a torpedo shot at "Lusitania," which was hit on the starboard side near the bridge. The detonation of the torpedo was immediately followed by another explosion of immense power. The ship quickly overturned to starboard and began to sink. The second explosion must be attributed to the ignition of the ammunition on the ship.

*The Deputy Chief of the Admiral's Staff
Behncke*

14

A trench leads from firing position 1/96 to the observation post.

When Reisiger arrives at the observation post, the captain was already standing at the scissor scope. The chalky wall of the trench is splattered with blood. Something is lying under a canvas sheet below. Reisiger reports in. The captain keeps his eyes on the glass and says: "We'll get rid of Raschke later. Put the phone to your ears."

Reisiger straps on the headphones and crouches down. Line test: The firing position answers. All right.

Everyone in the trench is on their feet. Man stands next to man. Everyone looks forward motionlessly. The rifles are inserted through the protective shields. There is no firing.

Mosel speaks now and then. "Hmm, hmm … well, well …" He talks to himself.

The company commander comes and greets him. They talk about the terrain. The company commander confirms the reports from the patrols that were out tonight: The enemy has undoubtedly cleared away the wire obstacles in several places. He had also heard this from the neighboring companies. The general opinion is that the attack will come in a short time.

That is enough. The captain decides to fire. Reisiger calls the firing position and gives commands.

"Battery ready to fire …"

"Fire!"

The shot sweeps over the observation. Crash!

Mosel is very thorough. The fire of the six guns is distributed in such a way that each gun is fixed on three different points of the enemy trench.

After a short time, the shots are right on target.

Reisiger can see nothing of the effect in his hole. But he hears the lively cheers of the infantry. "Hah … direct hit! Give him a good one!"

The enemy does not put up with this for long.

Reisiger picks up a report from the firing position: "Enemy firing. Apparently looking for the battery. But the shots are far behind it, about halfway up."

Suddenly, an infantry lieutenant arrives: "Herr Hauptmann, we've just received a report from the battalion that the attack is expected at 5:30."

Mosel leans over to Reisiger. "Ask how much ammunition is in the firing position."

Report from the battery: "About 400 rounds per gun!"

Mosel snatches the device out of Reisiger's hand, demands Lieutenant Busse and yells at him to make sure that one thousand rounds are procured for each gun as quickly as possible!

Then he checks his watch. It's 5:20.

The infantryman is still standing next to him. "Where did the report come from, Herr Leutnant?"

"Allegedly defectors, Herr Hauptmann."

Mosel points a finger to his forehead. "That means in ten minutes! Then the Lümmels have either cheated or they mean 5:30 this evening. Don't think the enemy is so stupid as to attack without artillery." The infantryman shrugs his shoulders.

5:30. Nothing happens. The enemy is a little more active with his artillery. Sometimes, it even hits the trench. But no sign of an attack.

Finally, the infantry lieutenant leaves in a huff. The alert is canceled. Only a few posts remain. The remaining infantrymen hide in the dugouts.

Mosel checks the section with the scissor scope again.

Boring.

"Reisiger, you sit down at the scissor scope. I'll go to the firing position. If anything happens, report it."

15

Looking through the scissor scope.

You see that all things that are far away from the naked eye have suddenly moved closer to you.

The enemy trench, a narrow white line in the terrain, a foot high, suddenly appears to you as a mountain range.

One mountain is separated from the other by a small valley.

In each valley is a black hole. A round, shiny something stares out of the hole: These are the muzzles of the enemy rifles. Sometimes, they jerk and flash. Nobody knows why.

In front of these mountains, you see a forest with huge trunks and sharply jagged branches that intertwine: wire obstacles.

When you turn the scissor scope to the right and left, you see that this forest stretches out immeasurably. Your eye finds no boundaries.

Sometimes, a trunk is bent, sometimes, the branches are torn off.

Sometimes, there is a hill in the forest. Bluish, set off in red. Oh, a Frenchman!

You'd think there was a person lying in the sun looking up at the blue May sky; his arms stretched out wide because the morning is beautiful, and the larks are singing. His two hands grasp the fresh clover. But no, these people do not see the sky and do not hear the larks. These people are fallen enemies. Three steps away from them, by the mountains, their brother stares through the black hole, occasionally striking the shiny something in front of them with his brother's hand. That's the only amusement they've had for weeks, for weeks.

You turn the scissor scope away from them. Aren't the living secrets behind the mountains easier to fathom than the silent ones in front of them?

Reisiger believes he has discovered living secrets. Behind the mountains, there is an incessant coming and going. Blue caps flit past, one after the other. You have to count them! But there are too many, far too many.

Reisiger presses his eyes hard against the glass.

Something else strange is happening, something inexplicable. The forest with the branches shifts in different places. Is this the fist of a giant that can move trees aside?

But you can't dream here! What is happening? For heaven's sake ... invisible hinges are used to carefully pull the wire entanglements closer and closer to the trench. Then, many hands reach over the parapet. And the wire entanglements disappear into the depths!

The blood pumps up in Reisiger. "Hello, battery, battery! Report to the captain immediately that the enemy is clearing wire entanglements. Hurry up!"

And the reply: "Captain is coming. Report to the infantry company commander. Over."

Mosel arrives with Sergeant Uhlig and Gunner Germer. "Reisiger and Germer, you take Raschke with you back to the position. Reisiger, you stand ready to relieve the sergeant in the evening."

A dead man is much heavier than a living one. And the heat . . . in the narrow trench. They have to lay Raschke down several times and catch their breath.

Behind the firing position is a dried-up stream. They put the body in there. They can take it to the Protzen later tonight.

16

Midday is approaching. The afternoon passes. Nothing happens.

Reisiger hears that all the comrades have been hauling ammunition since this morning. The columns could only get halfway up the hill. From there, every single basket—twenty-five pounds on each hand—had to be carried up. That was no fun!

"Here . . . both hands bloody . . . but the columns always have their pants full, cowardly bitches!"

They sit next to the guns, letting the sun shine on them.

Reisiger has to talk about the trench. "Yes, such a scissor scope is a fabulous invention . . . . Seen all the Frenchmen in person? Well, watch out, we'll soon have the most beautiful mess!"

The storytelling makes you tired. Maybe hunger too. The best way to get over it is to sleep. And bread is to be sent again toward evening.

Reisiger lies down and closes his eyes.

He can still hear Sergeant Gellhorn telling him that he is now to be a gunlayer again. And then he starts a half-loud conversation. Jaenisch, Hohorst, and Jahn are talking about things from home. In his half-dream, everything becomes so blurred that he finally no longer knows where he actually is. Then he sleeps.

Suddenly he is startled awake. Somewhere, there is a salvo of four shots. He wants to jump up. Gellhorn holds him by the knee, laughing: "Why so nervous? A heavy battery has come up behind us today; halfway up here, they're doing targeting drills now."

The first salvo is followed by the second, followed by many more, from all over. Jahn tells us that a total of eight batteries arrived last night, five light and three heavy: "The Frenchman better stay in the trench! Otherwise, we'll batter him until he loses his breath."

Reisiger stretches out. "It's the same with them as with us. What is ordered is done. They might have to attack, even if there are fifty batteries here."

Gellhorn deflects the conversation. "You're talking tin. You can't tell me anything. As soon as the weather gets better, all this nonsense will stop. Front march, march! And then we'll run over all those rickety dogs over there. So that this crap finally comes to an end."

The firing of the batteries has gradually increased to such an extent that you can hardly understand what your neighbor is saying. But you get used to the noise. It doesn't get on your nerves, but makes them vibrate, which is good for you. *This is us! Yes, that's us!*

Reisiger stands up and spreads his arms. "We can be proud of the fact that we are soldiers! I'm part of a battery that's right here in front of the enemy!"

He is about to sit down again when there is a hiss in the air. He sees that his comrades are moving closer together, half frightened, half curious. About eighty meters behind the firing position, four clouds of smoke spurt up. And at the same time, four barking detonations erupt.

The enemy is firing!

That changes the situation. The firing of your own batteries suddenly gets on your nerves. You curse. "I could have guessed that the stupid banging from the heavy battery behind us wouldn't go well in the long run," says Gellhorn, offended.

Reisiger wants to reply. Just then, a new hissing noise arrives, forcing him and his comrades to crawl close behind the gun. Crash! A moment later, the stench of a black cloud drifts across the grass. No one is swearing anymore. You have the impression of being trapped. You can no longer stand up because it seems nonsensical to raise your head above the protective shield. You can no longer joke because something is choking your throat. You try to talk. But that doesn't work either, because the hits are now so close together that you keep falling to the ground.

The batteries have stopped firing after the enemy's first salvos. That doesn't make him stop. On the contrary, you can see that two of his batteries are systematically searching the entire crest of the height. Soon, they must have caught 1/96. Closer! Further away! Further away! Closer! Closer again! Closer still!

Finally, four shots hit the right wing of the position.

Reisiger's back hurts. Gellhorn and Jahn had crawled very close to him and now the weight of their bodies was repeatedly resting heavily on his shoulders. Only when everyone lifts their heads to see the impact does he feel lighter.

Gellhorn then gives brief explanations. He speaks to them through clenched teeth. "Sixty meters behind us! Gosh, they're already dangerously close . . . Head away!"

Just as a group of shots rings out, a man jumps between Gellhorn and Reisiger. They glance up: Lieutenant von Stork. "If only we could fire too," he says in a trembling voice.

It has a calming effect. Everyone feels it would be a relief! It's almost a forced coupling. If the enemy fires, you have to be allowed to fire too. *The waiting is unbearable!* But the captain doesn't give the order.

The enemy has changed tactics in the meantime: He scatters randomly at rapid intervals, with lots of single shots.

One cloud of smoke after another dances around the position. In front and behind, closer and further, once so close that four birch trees in front of the left wing gun fly up, uprooted with a mighty roar, and lie groaning on their sides.

The minutes become more and more agonizing. If only we could fire too! At all the guns, the crews lie on their bellies, hands clutching the grass, waiting for the order to fire.

Finally, there is a distraction: a heavy battery, half left behind 1/96, takes up the battle.

The first salvo against the enemy! After two minutes, the next one follows. And shot after shot.

A salvation.

The enemy automatically concentrates his fire and puts it more on the slope, away from 1/96.

You can even raise your head. You can even kneel upright. Oh no, the fire is at the very back, by the column. You can even stand up.

Von Stork walks with emphatically slow steps to the telephone dugout. Busse is sitting there, the phone to his ear.

Seconds later comes the command: "Get the guns ready to fire!"

Now you're back to life. Now you know what you're there for. The first lieutenant's calm orders come from the telephone. The guns are aimed and loaded.

The battery fires the first volley.

Everyone's faces relax.

"Continue firing slowly at 2,800."

Thank God, they have something to do!

"Shots are good," shouts Busse.

"You should at least be able to see where you're shooting," says Reisiger.

"You'd better keep your mouth shut and aim properly," replies Gellhorn. "Everything else is none of your business."

Then the lieutenant arrives. "Reisiger, we've lost contact with the captain. Come on, mend the line . . . take Jordan with you. Gellhorn, you'll have to help yourself until then."

17

Reisiger sets off with Jordan. They have a telephone, wire, and insulating tape with them.

They duck their heads and run. The enemy's shots hiss over them. But you've gotten used to it. It's not too annoying.

The line is easy to follow. It is tucked between the branches of the small bushes in the foothills.

Everything is fine. Keep walking.

Then, they are exposed. Another ten meters to the approach trench.

Jordan runs ahead. As he jumps, he points with his hand: "Thick air. My God, it's being raked!" He's already in the ditch.

Reisiger stops for a moment. Damn! Unmistakable clouds of white and black smoke are billowing up ahead. Is the whole earth on fire? Horrible. But we must probably . . .

He leaps after Jordan in a single bound, staggering against the trench wall.

For now, they feel safe.

The wire is fastened here at shoulder height with small wooden stakes. It's easy to miss.

Then, the trench is blocked. A limestone pile prevents the passage.

Reisiger takes out the sidearm and digs. Jordan kneels down next to him. "There we have it. Shot to pieces. These oxen can't aim."

But the wire is whole. They both pick it up and let it slide through their hands. "'Cause not," Jordan grumbles. "So go on. Nonsense in this heat. Let's get on with it now . . . Boy, boy, it's smoking up ahead."

The trench makes various turns.

Finally, they find the damage. The wire hangs, torn with both ends on the ground. Jordan spits out, "There you go again. Not even shot to pieces. Some monkey from the infantry wanted to take it. Maybe to dry the laundry . . ."

They connect the telephone; it's humming. Thank God, the firing position answers. "Yes, this is line patrol."

Get the other end. Humming. "Hello, Observation 1/96." That's Uhlig's voice. "Herr Unteroffizier, stay on the line. We'll connect in a moment."

They want to twist the two wires together. They hear a beastly whistling sound diagonally in front of them. And as they throw themselves flat on their faces, a shot hits the bottom of the trench just in front of them.

Reisiger feels a shaking tremor in his limbs. He is freezing cold. He cannot move. Motionless, he must listen to the disgusting chirping of the explosives flying through the air. Then they hit the trench wall matter-of-factly.

But then Jordan pushes him. "If they carry on like that, they'll hit us with another one," he says. It's supposed to sound funny. But it doesn't. Jordan realizes it himself. "I think we're getting out of here. Leave the wire behind. We'll look for a shelter until the Frenchman has run his course."

Now Reisiger has courage again. "All this work will only take a few seconds. Go on, Jordan." He stands with his back to the enemy and connects the two ends of the wire. Buzzing. "Hello, Battery!" Once more, damn it. No response. "Hello, hello, Battery!!!"

Off. So, another point must have been disconnected in the meantime.

Jordan snatches the phone from Reisiger. "It's bone chilling. Let me have a go. Or shall we call Uhlig again first? Hello, Observation 1/96?"

An overloud voice rings out of the receiver. It's the captain. "Herr Hauptmann, this is line patrol."

"Get over here!" yells Mosel. "Over here now! Stop the stupid patching up, got it? But get over here now."

They pull the device off the wire and look at each other.

"That's all we needed, now," says Jordan.

Reisiger shrugs and takes the cable reel under his arm. "It doesn't help."

They run off.

After a few steps, a shot lands on the trench wall just in front of them. Reisiger looks down and sees that a large piece has been torn out of the limestone floor. The rubble almost fell over them. They jump up and over it. Then they are lying down again. And for the next fifty or sixty steps they crawl on their stomachs. As far as they can see, yellowish fires and black clouds erupt. The crashing is so loud that it hurts their ears. Again and again, they leap forward. Again and again, they lie down, their hands and faces pressed as hard as they can into the limestone.

At last, they see the front line. Twenty more steps! There's the captain! No longer at the telescope, he crouches half bent over next to the hole in which Uhlig is sitting. Sometimes, the black of the clouds of smoke obscures him.

Once, just as Reisiger is about to jump again, he disappears in a flash. But when Reisiger looks up again, he is still crouching and motionless in the same position. He waves, puts his hands to his mouth, and screams. You can't understand it.

They reach him with the next movement. Reisiger tries to stand at attention and make a proper report. Suddenly, he flies into the captain. Their heads collide. Then they are lying in the telephone hole, Uhlig below them and Jordan above them. As they finally disentangle themselves, Mosel shouts angrily and pleadingly with all the strength in his lungs: "Run back, tell the battery to fire rapid fire twenty-six hundred! Tell Oberleutnant Busse to try to set relais!"

Back? Now? There's nothing to say.

Reisiger sees the infantry, shoulder to shoulder, rifles at the ready, between smoke and fire.

Jordan is already chasing ahead of him. He follows. Throwing himself down, a short jump, throwing himself down again, crawling on his stomach, fanned by hot flames, spit by acrid smoke, pours of lime and dirt above him. The battery at last! Here, there are no more forms and no more reflection on military posture. Reisiger leaps at the first lieutenant: "Rapid fire twenty-six hundred!" Then he chases past Busse to his gun, repeats half-madly and completely out of breath: "Rapid fire twenty-six hundred!" and pushes Gellhorn away from the gunner's seat.

And already, like the signal of a blaring trumpet, Busse's voice: "Whole battery, rapid fire twenty-six hundred!"

18

The battery becomes a machine.

On the command "Rapid fire twenty-six hundred," six men on six guns place six shells with the fuses set at a range of twenty-six hundred.

On the command "Rapid fire," six gunners aim six guns at the range: twenty-six hundred.

The "rapid fire" command causes the arms of six gunners in a battery of six guns to tear open the breech six times in sixty seconds with the precision of six machine levers, six other gunners to insert six shells into six barrels six times, six breechblocks to be thrown shut at the same second, six right hands to detonate six shots six times at twenty-six hundred. Six barrels then recoil six times in sixty seconds so that the guns rear up like beaten animals.

It is loaded; it is aimed; it is fired.

"Rapid fire" means that after ten minutes, people's pulse rate is doubled. The heart no longer beats in the chest but in the throat. First, the pulse makes their limbs tremble. Then they brace themselves against a command, become like iron and become part of the great machine: six guns, a battery.

"Rapid fire" means that after half an hour, the crews of six guns tear open their skirts with automatic movement. After one hour, some crews take off their skirts, some open their shirts, others roll up their sleeves.

"Rapid fire" means that after an hour the crews of the battery have the pallor of death on their faces, over which soot and powdery slime smear a thick black.

"Rapid fire": People sometimes try to shout something to each other. After a short time, each attempt is silenced. When the attempt starts again, each shout becomes an animalistic scream.

"Rapid fire": The men's fury is channeled into the guns. Six cold metal barrels sound the death knell six times in sixty seconds. After a short time, they hiss whitish steam, sweating like the people, the people working on the machine. Then the machine thirsts for blood: The barrels are as hot as a fever.

"Rapid fire": The fever becomes a contagious disease. The fever poisons the soil. Once the ground was green, the first spring green. After sixty minutes, the green is crushed, trampled, ground up. Six times, the earth has two deep wounds in which the gun wheels mercilessly cut six times in sixty seconds.

It is loaded; it is aimed; it is fired.

Loaded aimed fired.

19

The First Battery F.A.R. 96 has been firing rapid fire at 2,600 for an hour and a half.

All the batteries on the crest of the height are firing like this. The order "Rapid fire" must have reached them all. Foot artillery shot after shot, field artillery shot after shot, over to the enemy.

It is impossible to tell whether the enemy is responding.

At one point, Reisiger feels a hot breeze close above him. Is it coming from over there? He wants to look around, look for the impact, but there is no time. Rapid fire, rapid fire! Gellhorn looks anxiously to his right and left. Ammunition? The empty baskets are piling up. But for the time being, there is no shortage.

Just as the sharp smoke from the muzzle flashes back and the gun rears up, a group of birch trees shakes halfway to the left. Reisiger sees that the whole bunch flies up and lifts the roots against the sky. So, after all, the enemy is firing too. The shot was ten meters in front of us.

He wants to draw Gellhorn's attention to it. He doesn't get the chance. Von Stork suddenly stands next to him and shouts in a voice that has nothing human about it: "They're through! Set fire to eighteen hundred!"

They are through? Eighteen hundred?

There is a shrill cry. And as the range is set to 1,800, and as the ammunition gunners load the first shell with 1,800, they all see that

Lieutenant von Stork is lying bent over on the ground, his right gaiter torn open. A stream of blood gushes from his knee.

Rapid fire! "But someone has to help," Reisiger says to Gellhorn. But he has long since taken a new grenade himself and pushes it into the barrel. Come on, rapid fire!

What's going on now? The two right wing guns suddenly changed direction, made a quarter turn. What does that mean?

A voice shouts ... we don't know from whom or where: "The enemy is firing from the right with machine guns into our flank!"

Doesn't Gellhorn hear the shout? Reisiger wants to draw his attention to it. Then he sees First Lieutenant Busse jump to the lieutenant. Does he want to bandage him? Jesus. He throws both arms in the air and falls on his back, wincing.

"Herr Unteroffizier ..."

"Rapid fire eighteen hundred, go, go!"

Rapid fire on 1,800. The gun is now so over-shot that the recoil fails. Reisiger and Hohorst have to push the red-hot barrel back into place after each shot.

But otherwise, all movements remain mechanical, rapid fire!

Gellhorn has now seen the lieutenant. He is hunched over, holding the lid of his cap in front of his knee. The blood drips more slowly. Busse lies still, no longer moving.

"Boys, the battery is without an officer," says Gellhorn.

The captain is standing in the position like an apparition.

Everything pauses for a moment. All eyes are on him.

He has a chalk-white, motionless face.

"Rapid fire!"

He walks upright from gun to gun. "Rapid fire, guys!" To Gellhorn he says: "Shoot! At sixteen hundred! And if the enemy is there, we'll have to see how we get out of here."

Rapid fire.

After a few shots, there is a roar like Reisiger has never heard before.

He realizes that he is sliding off his seat.

Then, it is completely dark.

As the darkness lifts again, Reisiger dreams. He senses that he is lying in a lake. The water in the lake is very warm. Can't you swim here? He has stretched out his legs and arms, but he can't swim because he is weighed down by a burden.

The waves of the lake beat loudly against the shore.

These crashing waves become more and more violent, more and more threatening! And finally, scare him. *I have to swim; otherwise, I'll be crushed!*

He lifts his chest and pushes his elbows against the bottom of the lake.

He is suddenly awake and opens his eyes. He hears a terrible crash on all sides and feels a pressure above him.

And as he lifts his eyes, he sees sharp spikes everywhere in front of him. And as he lifts his eyes higher, he sees that his gun has collapsed on one axis above him. The barrel is smashed beyond recognition. And as he searches for his comrades, he finds Gellhorn beside him without his head and Hohorst with his arms torn off. And behind him lies a single leg. That is all. He carefully pushes the sergeant away from him and kneels down. He looks to his right. Two guns are firing rapid fire. He looks to the left. A gun is firing. He wants to get up. But he falls down again. He grits his teeth and tries a second time.

The captain has noticed him, jumps to him and says: "Go down and look for the Protzen. But it may already be too late. Battery, rapid fire nine hundred!"

Reisiger runs down the height. But when he is barely a hundred meters away from the battery, any attempt to reach the Protzen seems completely pointless. The huge oaks to the right and left of the paths have been battered, and there are cracks everywhere, and more trunks are breaking. Fountains of enormous lumps of earth are hurled against the sky. Wherever you look, fiery flames are shooting out of the ground.

Reisiger tries every possible way to move forward. He crawls on his belly, pushing himself from crater to crater. He jumps and ducks behind tree trunks. He jumps up, only to crawl again after a few steps.

Sometimes, he stays down and closes his eyes. *If only it would finally strike me!*

Until, the next shot howls next to him and chases him on.

Finally, he reaches the bottom of the hill. A barrage of bullets roars to the left, burning houses to the right. *Maybe*, he thinks, *I'll find the Protzen there*. So, he keeps to the right. As he stops once again, panting, with a black haze before his eyes, the thought *Protzen* becomes more insistent: *I must get the Protzen; otherwise, the whole battery is lost!* Picks himself up and collapses. Up again. The next moment, he is torn through the air by a shot directly behind him. Then he smashes his skull into the water.

He remains motionless there for a moment. Now he's getting cooler, and that's good. The pressure lifts from his forehead. He suddenly realizes again what he is looking for down here. *I've got to get the Protzen! Otherwise, the whole battery will be lost!* He says this half aloud to himself. He wonders how he is talking to himself. He laughs for a moment. Then he gets up and walks on.

Close to the burning houses, he sees Sergeant Hollert clinging to a smoking wall. He rushes up to him and shouts, "Protzen forward!" Hollert is startled. Reisiger adds, "Herr Wachtmeister, everyone from the Third Gun is dead except me." Then he collapses. He hears the gallop of horses and the rattle of wagons next to him.

20

Report from Dr. Dülberg to F.A.R. 96:

In the village of Farbus yesterday, May 11, 7:15 in the morning, I found the gunner Adolf Reisiger 1/96. Identity from pay book and identification tag.

I noticed: Skirt torn in front, bloody. On chest palm-sized, bloody, suggillation spot on right third to eighth rib. Bullet hole not present. Blood clotted on mouth and nose. Pulse small, rate 130, unconscious.

R. was lying in front of a burning hospital ward, which was deserted because heavy artillery fire had forced the evacuation of the village. I took R. on my horse, but later, as my horse was shot down, I handed him over to a passing ambulance car to be taken to the field hospital.

21

Report from 1/96 to Regiment:

Casualties on May 11: Dead: two officers, three non-commissioned officers, eleven men. Wounded: four men. Missing: one man.

22

The enemy in front of us seems to have only a few divisions. We are four times as strong as he is and have artillery more formidable than has ever appeared on a battlefield. Today, it is no longer a question of daring a coup de main or taking a trench. It is a question of defeating the enemy. That is why we must attack him with the utmost ferocity and pursue him with an incomparably tenacious fury, without worrying about fatigue, hunger, thirst, or suffering. Nothing is achieved if the enemy is not finally defeated. So let everyone—officers, non-commissioned officers, and soldiers—be convinced that from the moment the order to attack is given until final success is achieved, France demands every boldness, every effort, and every sacrifice from us.

The General in Command of the XXXIII Army Corps, Petain

Chapter IV

1

Night has fallen.

The battle continued. The troops in the quarters behind the front line have been roused. Soldiers are loitering on all the village streets, standing together, silent, dull. Sleep is out of the question.

The horizon, with its flaming whitish wings, is a spell that turns everything into a stubborn torment. The backdrop of a stage on which ghosts act.

In some villages, the spectators suddenly become players. The guards run at a trot, roam the groups of curious onlookers with hoarse shouts, and shout in every quarter. And then it's swarming like an anthill.

"Get ready to march."

The call is heard mainly by the infantry. It stifles even the faintest whisper. You curse under your breath. Then, they go silently into the shelter and take their rifles, cartridge pouches, and helmets. After a few minutes, the companies are in place. Then, the march begins silently.

Silently, the officers at the head of the columns, silently the non-commissioned officers on the right, silently the whole group.

There are no illusions.

There is no point, no point at all, in even imagining the next hour, let alone the next morning. For every thousand men who cross the road, half of them must know that they will be torn and shattered by morning. But they don't think about that.

Each man is relieved of his own responsibility by the command: "Company march!"

Up in the distance, the heights are burning. The order is: Extinguish it!

Up in the distance, the brothers are shouting. The order is: You have to help.

Up on the heights, the enemy has broken through. The order is: They must be driven back.

The command. That is all.

Dull and dull and half asleep, the columns crawl forward.

Their minds only brighten when other columns come toward them.

A car drives past, with a fluttering map, the Red Cross on either side. Many cars, ten, twenty in a row.

They push past very carefully.

And when they have disappeared, a lifeless body lies somewhere, against one of the trees on the highway—left lying there because it would have been pointless to take it backward and because it would have taken up space for others.

And further on.

Now, soon in the firing range of the enemy artillery, the wounded are swarming.

Since the ambulance cars can't get here and since there are no doctors, you have to help yourself.

One man after the other.

Put away your weapons, put down the paddock, use a club as a support . . . "Just get home."

"Just get out of this mess."

"Get home."

"Get home."

"Get home!"

And I'd rather crawl on all fours than stay here.

Onward with the columns, to the front!

You're often tempted to ask one of the wounded something. It would be good to know how the battle is going up there.

Onward with the columns.

Nobody asks; we'll find out much too soon.

2

On the edge of a village, near the fire zone, lies a blooming orchard. It smells of spring.

Underneath the blossoms of the trees are the remains of the First Battery F.A.R. 96, four mounted guns and two guns without mounts.

There are seventeen horses, ten gunners, six drivers, three non-commissioned officers, and two officers gone.

What is still alive is huddled together. The gunners sit, leaning back to back. Occasionally, one hangs his head and tries to sleep. They all want to sleep. But when the fires light up the garden and the horses tear up the harness, sleep is shattered again and again.

And people are hungry. But there is nothing to eat.

An hour ago, the captain sent a sergeant into the village to get bread from somewhere. But the sergeant hasn't returned yet.

At least you want to smoke!

There is one cigarette in the whole battery. It belongs to Gunner Lechter. As he lights it behind his cap, the captain stands next to him and says, "Give me a puff." He presses his hand on his shoulder so that he can't stand up. "I'll give you a whole pack as soon as we get mail again."

Lechter gives the captain the cigarette. Mosel takes a puff. Then he hands it back. Lechter is the second to take a drag. Then, the stub makes the rounds among the next comrades. A sergeant and five men join in. Then the fun is over.

Mosel goes to Sergeant Hollert. The night is oppressive. You have to talk to someone. Hollert pulls himself together, but then he realizes he can hardly move his limbs from fatigue.

"Please stand comfortably, Sergeant. Or come on, let's sit down."

They attempt a conversation. They have the feeling that something needs to be said about the past day. But where to start? For the time being, it remains a matter-of-fact discussion.

"I don't understand where the sergeant is with the rations. Have we run out of iron rations, Constable?"

"Most of them have lost their luggage, Herr Hauptmann."

Mosel shouts, "Those who still have iron rations can use them up!" After a while: "Do you think, Sergeant, that we can pick up some mail tomorrow? I'd like to make people happy."

"By your command, Herr Hauptmann. It's been almost eight days since the last issue."

Again, after a while: "The most important thing is to get two new guns quickly. I don't understand why the regiment hasn't sent an order for them. Sergeant, we have to send a dispatch rider out first thing in the morning."

"By your command, Herr Hauptmann."

Mosel and Hollert keep looking at the horizon. It's burning, it's burning, and it's roaring incessantly.

Finally Hollert says what he's thinking: "We have relatively heavy losses."

"Well, Sergeant . . . that's how it is."

"Will Herr Hauptmann allow us to fetch the dead if we're still here tomorrow? The Second Gun wanted to put the First Lieutenant on the Protze. But it held up too long yesterday."

"Sergeant, what good is that? Leave them where they are. You've got nothing left of it . . . Tell me, where has Reisiger gone?"

"I saw him fall over when I gave the order for the Protzen to gallop. We reported him missing. But he's dead. He already looked half-dead when he came to us . . . After all, the battery might have been lost without him."

"Tell me, Sergeant, how did you get through to us?"

"How you get through something like that, Herr Hauptmann . . . I don't know myself. There wasn't any fire like that in '14. We only drove through hail."

"Yes, yes."

3

Reisiger felt a prick in his arm. It woke him up. It was daylight. He was lying down. Two people were standing in front of him. One smiled and said, "There you go." Then he turned away from Reisiger to his neighbor, a man with a big, black mustache, and said to him, "So it's more harmless than we thought. Give him ten drops at lunchtime and ten drops in the evening. And we'll talk to him in the morning."

Then they both walked away from Reisiger. He closed his eyes again.

After a while, he woke up again. He couldn't find his way around. Finally, he realized that he was lying in bed.

In bed? He raised his head slightly, startled. A blue plaid feather blanket was billowing out. He threw his body up a little and let it fall. Yes, it was true; he was lying in bed. He looked to his left. The bed is against the wall of a room. Looked to the right. The room is large and has bright windows. There are four more beds between the windows. There are also people lying in them.

Reisiger finally straightened up and sat down, despite a tugging pain in his chest. He cleared his throat. He wanted to make himself heard because he didn't know what to say.

Clearing his throat worked. Four people rose from the beds. They looked at him. Then one of them said, "You're surprised you're alive, Comrade?"

Reisiger had to laugh. "We're in a military hospital here?" he asked.

"You're a fabulous guesser," he grunted.

"But I don't know what's wrong with me."

"But we know exactly what's wrong."

He was told with medical thoroughness what was wrong with him. He had been shot in the chest. The shrapnel must have bounced off because his chest was not shattered. Instead, he suffered a paralysis on the left side. It will be resolved in a few days.

Paralysis? Reisiger was shocked. He tried to lift his left leg. No, he couldn't do it. Then he tried to stroke his hair with his left hand; that also went wrong. He couldn't lift his hand.

"Why do you know better than me?" he asked.

The man who had greeted him first, apparently the spokesman in the room, clarified: "I've been in this dump for six weeks. I know the hustle and bustle. And when the little doctor makes his rounds with the sergeant in the morning, he's always very busy. He wants to show us everything he knows. The result is always the same: 'Sergeant, you have to brush with iodine.' The other day, a guy came in without his head, and he had him painted with iodine until he could walk again." He roared with laughter at his joke.

Reisiger had noticed the Bavarian dialect in which he was being spoken to. He asked, "You must be Bavarian, Comrades?"

The spokesman put his hands behind his back. "You bet we're Bavarians! This military hospital is our Bavarian property, and no carrion has any business in there. Or are you not Bavarian?"

Reisiger was not prepared for the question. What is this supposed to be doing here in a war and among German soldiers and in a military hospital and reasonably close to the front? He didn't really understand. He answered with indifferent matter-of-factness: "I'm a Prussian."

Was that wrong? he thought. *Shouldn't you say something like that?*

His answer had a strange effect. The occupants of the four beds threw themselves back into their mattresses as if on command. Reisiger no longer existed for them. He asked, "Do you mind?" No one answered. He tried a long explanation that he didn't understand why anyone would object and that they were all comrades after all. No one responded to that either.

Finally, he gave up his efforts.

One Prussian bed against four Bavarian ones—you can only lay down your arms.

A man in a drill suit appeared in the room. He was carrying four enameled tin bowls, which he distributed to the four Bavarian beds. "We're having bean soup today," he said. The four of them tucked in, slurping loudly. The paramedic wanted to leave. Then he changed his mind and turned to Reisiger. Where did he put his cooking utensils?

Reisiger shrugged his shoulders. "I don't have any." He explained how he had run down from the heights without any luggage to fetch the Protzen. That made no impression. The medic had no sympathy for forgetting the cooking utensils. And when the spokesman for the Bavarian beds interjected a remark as to why the bloody Prussians were being fed here at all, he probably felt like leaving Reisiger without food. But then he looked at him scrutinizingly, mumbled a few unintelligible words, not entirely unfriendly in tone, and left. And when he came back, he brought a rusty, crookedly cut tin can and a tin spoon with the handle chewed off. "Eat it."

The meal was exhausting. Reisiger clamped the can against his chest with his left arm. It was very painful.

He felt depressed.

It's so unpleasant to lie here, useless, betrayed by his body. And this disgusting atmosphere in the room.

By the evening, the paramedic had also passed on. He now clearly took sides with the Bavarian compatriots. He went so far as to respond to Reisiger's requests to be allowed to wash himself by saying that the Prussians could walk around as dirty as they liked. The Bavarians had to do all the work to win the war, and that was why the Bavarian wounded would have their turn first, and then the Prussian wounded would not have their turn for a long time.

Reisiger lay awake that night. It was very quiet in the house.

And outside? It's strange how far the sounds of the last few days have moved away from here. In the far distance, there is a humming in the air; sometimes, the windowpanes shake. Otherwise, there is great silence.

The silence is oppressive. The memory of the battle suddenly emerges from it. It torments him. He can't put any details together and can't organize his thoughts because images are coming out of the darkness everywhere, so fast, so rushed, so confusing. He sees gun barrels, their fire ricocheting around. Then impacts splash up. Then faces appear. One laughs with white teeth. *Who is that? Who is that? What was that?* Yes! It had been black around him, and when it got light, Sergeant Gellhorn was lying next to him with his head torn off. And one leg was lying there, wearing one of Hohnert's new boots. And Hohnert? And the gun had a broken barrel and a shattered wheel.

Reisiger struggled to sleep. He tossed and turned. When he failed to fall asleep, he was so angry that his tongue tasted bitter.

Face to the wall: Why this damned filth!

Face to the window: To desecrate this beautiful night like this!

Facing the ceiling: Why ... have ... you ... left ... us!

And back and forth with accusations and doubts. *Tears in my eyes, acid in my throat.*

Until it finally got a little brighter.

Then, the feelings changed. With a soft hand came the awareness: *I, Adolf Reisiger, am in the military hospital, and I am safe.*

But "safe" was a word that created new unrest.

Reisiger calculated that apart from the two officers and the operator of the Third Gun, more people from his battery had certainly fallen. *So now*, he thought, *this battery is standing around somewhere.* The Third Gun is missing, two officers are missing, six or eight men are missing, and the battery is no longer ready to fire, and so needs every man! So, being safe means taking a man away from a battery in need!

He gripped his left hand with his right. It was difficult to move it. *Is that a reason to be safe here?* He felt his chest. It hurt a lot. But that didn't excuse it either!

And then, with a rapid heartbeat, the decision came to him: *I must go to my battery immediately.*

At first, a doubt fought against it: *What's one man more or less?* But the doubt won out, and the desire grew stronger: *I must get to my battery!*

Reisiger straightened up. *Where is my uniform?* He looked all around the room, but on a box in front of him was nothing but a blue-and-white striped hospital gown.

But that didn't discourage him: "If I put on the stuff and sneak out of the room, I'm sure I'll find a uniform somewhere. It doesn't have to be mine. I'll take what I can get. I've got to get to my battery!"

He threw back the covers, his right leg already on the ground. But as he pulled his left leg back with his hand and tried to stand up, he buckled and slumped back. He tried a second time. Then he gave up. The attempt had made him so weak that he could barely get his legs back under the blanket. *I have to get to my battery! I have to get to my battery!*

He lay down on his stomach and cried.

4

The battle of one Prussian bed against four Bavarian ones became increasingly hopeless. The numerical superiority was reinforced from all sides. Just twenty-four hours after his admission, Reisiger was regarded as a mangy dog throughout the house, which consisted of fifty occupied beds.

Not that he was scolded; much worse happened. He was completely negated.|

This went as far as the doctor, who never said anything other than, "Well, our Prussian will get well by himself."

And not even iodine was prescribed.

Finally, salvation came.

One afternoon, the door to the room slowly opened, and a very well-fed, clean-shaven gentleman entered. He was wearing a long gray skirt with purple lapels and a silver chain with a crucifix around his neck. "Greetings to God, dear comrades."

Reisiger looked up. The divisional priest. Another Bavarian!

The priest went from bed to bed. From the inside pocket of his skirt, he pulled out a leaflet, which he pressed into the hands of the wounded. He said the same sentence with regularity, "Comrade, it's getting much better, isn't it? You just have to be patient . . . in a very short time you'll be back at the front."

He had finally seen Reisiger. He approached him with measured steps. Then, he sat down on the edge of the bed and began an interrogation.

"Where were you born, my son?"

Reisiger gave his place of birth. "Father, you won't know the place. It only has a few thousand inhabitants. It's in the province of Saxony."

What now? The occupants of the Bavarian beds suddenly straightened up and grinned. Reisiger looked puzzled.

"So you're Saxon?" asked the priest.

I have to get to my battery, Reisiger thought. *I can only get to my battery if the doctor takes care of me. The doctor will only look after me if I'm not Prussian. Maybe the Saxons are better people and even appeal to Bavarians.*

He forced a smile on his face and said very loudly, "At your command, Father. I am Saxon."

The priest's tone became even softer than before. He stroked Reisiger's head with his hand. Then he pulled out the print from his skirt, placed it on the bedspread, and said, "Take a good rest; soon we will all be facing the enemy again, and rest assured, we will have won the victory in no time."

He waved his hand several times and disappeared.

From then on, there was an intimate friendship between the four Bavarians and the Prussian-Saxon bed. The spokesman for the Bavarians jumped up and went to Reisiger. "Comrade, you could have said that straight away. We Bavarians and you Saxons have always been good friends." He shook his hand. The others also greeted their newly discovered brother-in-arms.

The food carrier arrived. The Bavarian spokesman said with cold matter-of-factness: "That one over there isn't a Saupreussen at all." As a result, the crookedly torn tin can was immediately exchanged for a white enameled bowl.

The friendship became even closer. Over the day, the four Bavarians offered up their life stories for Reisiger in Bavarian detail, including the fates of all their wives and children, bribed the paramedic in the evening to fetch beer from a canteen, and drank on Reisiger's health until their mouths dripped.

The miracle of Bavarian-Prussian-Saxon fraternization blossomed in the most lavish manner. Firstly, Reisiger slept through the night without being tormented by thoughts because of the unaccustomed beer consumption. Secondly, the doctor took care of him the next day by prescribing hot compresses for his arm and leg. Thirdly, the Bavarian spokesman put the crown on the new alliance, namely, today was Saturday; Saturday the doctors were sitting in the casino, so all the wounded climbed over the wall of the hospital garden at night and went to the ladies of the village. Reisiger was cordially invited to take part in this excursion. He did not have to pay anything for it.

Reisiger declined the invitation.

I must go to my battery, and if I can go, every step must be a step toward my battery. He did not say that aloud.

But even so, he did not understand. "You're an ox, Comrade; we're allowed to fuck for free here in the village."

5

Protest against the undignified behavior of German women:

A protest is being raised on behalf of millions of German women against the despicable behavior of German women (or wives) who have crowded up to captured enemies at train stations and presented them with chocolates, roses, and other "gifts of love." This is nothing less than treason against the Fatherland. Betrayal of our good German reputation and name. The German authorities should act with the utmost severity.

As a girl of nineteen, I experienced the war of 1864, and I had a husband and brothers in the field during the wars of 1866 and 1870. Already in 1870, we had to witness the abominable spectacle of women who did not know how to preserve their dignity towards the captured French and Turkomans. That is why all decent women are now asking for the most ruthless action to be taken against women who display such undignified behavior.

Johanna Baroness von Grabow
Widow of Colonel von Grabow
(*Berliner Tageblatt*, August 18, 1914)

6

After two days, Reisiger was given a clean bill of health by the field doctor at his insistent request and released to join the troops.

None of the Bavarians understood why he was so eager to return to the front so quickly. He still couldn't understand why they, who had recently climbed over the high garden wall with elegance and ease, were still lying in their beds.

After all, his relationship with them had become so friendly that he said goodbye to them warmly.

He was given his uniform. There was a large bloodstain on the back. At first, he tried to wash it out, but then he didn't. He remembered that this blood came from one of his comrades, whose bodies had formed the cover under which he was allowed to live.

He was ordered to march to Douai and report to the stage commandant's office. The location of his regiment was unknown, but he would be traced there.

Why walk? Douai is two days' march from here. Is there no rail connection?

But you are not allowed to ask. Everything in the military has its well-considered reasons.

Reisiger marched off.

The village was soon behind him. It was beautiful May weather. The sky was blue. To the right and left of the road, the corn was as tall as a hand.

He was happier than he had ever been in his life. He had never felt so unburdened.

What can happen to me? There are no worries. All people must be good, because all people are German soldiers and comrades. You can lie down and sleep: you know you will wake up as if you were in the safest of homes, not robbed, not attacked, as happened on country roads in peace. There is no need for food. Because in every town there are canteens where you can buy cheap food, or field kitchens to which you are invited.

Comrades joined us for some stretches of the journey.

How noncommittal such encounters are. You don't introduce yourself; you don't pay compliments. You talk as you please, share your thoughts and wishes. And you part—you will never see each other again—always with the feeling that we belong together, we are soldiers.

A few times the field police came, riders in splendid uniforms with silver shields on their chests. They were looking for shirkers. But Reisiger had his identification card. He was safe with it.

On the first evening, he found a dilapidated wooden hut in the field. There was some straw inside. He slept until dawn.

When the sun went down again, he was in Douai.

7

On Tuesday at noon, the Italian ambassador in Berlin, Bollati, who had previously been left without instructions from Rome, received instructions by telegraph to request his passports from the Foreign Office too. He immediately carried out this order, and the passports were delivered to him in the course of the afternoon. This means that also Italy has now broken off diplomatic relations with Germany.

(*Leipziger Neueste Nachrichten*, May 26, 1915)

Chapter V

1

Meyers Konversations-Lexikon 6th edition 1909:
Lens: town in the French Depart. Pas-de-Calais, Arrondiss. Béthune, on the Deule and the canal of L. (part of the course of the Deule), junction of the northern railroad, has coal mines, manufacture of beet sugar, machines, rope goods, etc., and (1901) 24,370 inhabitants.

Der Kleine Brockhaus, 1925:
Lens: in the French Depart. Pas-de-Calais, 32,000 inhabitants., (coal) industry, Destroyed in the World War (rebuilt).

Der Kleine Brockhaus, 1925:
Loretto Height, altitude 165, hill in northern France, north of Arras with chapel. May 9 to July 27, 1915, focal point of the fighting in the battle of La Bassée and Arras.

2

Field Artillery Regiment 96 is called into action on a sultry, rainy June night in the year 1915, traveling along three roads. At one o'clock in the morning, six batteries with the horses of their First Guns stand against the massif of a mighty church.

The night is black. All voices are a whisper. All you can hear is the clanking of harnesses, here and there, a shout and half-loud commands.

All the officers are ordered into the church. A few candles are burning behind the heavy doors and at the altar. There, the regimental commander gives orders for the firing positions.

This takes less than half an hour. The regimental commander disappears. The church is empty.

At dawn, the six batteries of F.A.R. 96 are positioned in six suburbs of the town of Lens.

When day breaks, the gunners of the six batteries see a steep hill and on it a chapel with a glowing cross: the Loretto Hill, altitude 165, with the chapel "Notre Dame de Lorette."

3

The First Battery's main direction is toward Loos, a quarter turn to the right from Height 165.

The houses in the suburbs of Lens are uniform: red brick, two-story buildings lined up next to each other in colonies. They all have a wide entrance on the same side with a door painted green, two windows next to it and two windows above it. Behind the house is a courtyard of the same size, six or eight steps square. Behind each courtyard is a small stone house: the laundry room.

On one street, the doors of all these laundry rooms are forcibly widened. The 1/96 guns stand in the laundry rooms, their muzzles looking through wide windows.

The gunners take a childlike delight in this strange firing position. They are all eager to fire soon. How delightful it must be to politely open the windows before the first shot, so that shells or shrapnel don't damage the window cross or even the window panes, for God's sake.

The living quarters for the battery are as comfortable as the gun emplacements.

There are civilian living quarters with a sofa, table, chairs, real beds, and long-awaited, albeit dirty, bed linen.

There is one house for every gun commander and five men. *Is there a better way to live?*

The guns are barely positioned in the laundry rooms when they start inspecting the gardens in front. A few flowers are blooming, young bulbs, and the fruit trees have set well.

And then you go for a walk in the street. You put your hands in your pockets and stroll along the pavement. It's a pity that civilians are missing. Women. That would be cozy.

But is there a better way to live?

Admittedly, if you look to the side, there is the black height of 165, dabbed with white balls, clouds of shrapnel that never go away. And from time to time, the battery guard yells, "Cover, Airmen!"

But what does that mean? The war is a joy here!

What lies behind is over!

There's a full crew again, a full six guns!

Adolf Reisiger belongs to the Third Gun again.

The morning after his arrival in Douai, he had already been assigned to the battery.

It had been hard to get used to the fact that "his" Third Gun no longer existed. Certainly, there was a brand-new gun carriage on which "3rd Gun" had been painted.

But this stencil remained the only reminder.

Everything else was new. The gun commander, Sergeant Leonhard, had been transferred from the Third Battery; the gunners came from the garrison. Two, Aufricht and Heynze, were volunteers from a replacement transport from January. Rabs and Gunner Georgi, reservists, came from the recruit depot.

Reisiger had trouble getting along with the new people in the first few days. After all, it's not enough to share day and night, table and bed, gun carriage and Protze, sky, clouds, and earth. You have to find connections outside of outward appearances.

Aufricht is a theology student, Heynze a bank clerk, Georgi a carpenter, and Rabs a dental technician.

Gradually, on the marches of the previous nights, Reisiger, who sat alternately on Protze and the gun carriage, found access to the new comrades in short conversations, in jokes, in silence.

And they were already a good unit.

Rabs is particularly likable. A reserved man, a little older than Reisiger, with clever, clear eyes. It was a matter of course that the two of them slept in the same bed here in the house, that they tried to keep to each other as much as possible in all services.

And in the evenings, when they sat together around the table in the courtyard after dinner, they got up to mischief, in which everyone, including Leonhard, took part enthusiastically.

They played cards and dice. And, as a specialty, they played "drawing time."

Because Aufricht was an artist. He skillfully copied Wennerberg's watercolors, which were the main decoration everywhere in dugouts and trenches: girls, girls, standing, sitting, lying down. With flowing skirts, with long arms, with overlong legs. The longing of so many soldiers.

That's what Aufricht drew. And when asked, he did more: he undressed each of the ladies. On request, you could have the lady hanging on the wall in front of you naked within a few minutes. It was great fun, and the fun was intense.

If you were on good terms with Aufricht, he would give you one of these sheets. Or anyone who gave him a blank sheet of paper was allowed to pick it up again, along with a naked girl, in exchange for a cigarette.

Almost every night passed with this game. And the later the hour, the more naked the ladies became. Until finally, around midnight, they were doing the same thing in all the drawings, clearly, unambiguously: lying on their backs, arms outstretched, legs bent at the knees, thighs spread wide. In every drawing.

One such sheet would then cost three cigarettes.

4

At night, it was so warm that it was difficult to sleep. You lay awake, often until it got light again, tossing and turning, the sultriness oppressive. Thoughts crawled, sticky snails: What . . . should . . . we . . . do . . . here? Boring. Is . . . there . . . no . . . war . . . here . . . at all?

You felt so superfluous. The enemy wasn't firing—the battery wasn't firing. *So, what was it all for?*

And when you lay awake like that, in the fat, thick air that didn't let up even at night, then sometimes scraps of music drifted in, hurdy-gurdy probably, orchestrion, piano.

Hell, that must have come from Lens. "Comrade, listen, there's music in Lens. Do they dance there?"

"Well, gosh, it's a waltz, for crying out loud!"

"That's all we need. And they'll keep all the women at bay."

"Well, Aufricht drew you one yesterday, such a sturdy one . . ."

"You big git, shut up."

Music . . . and a big undestroyed town ten minutes away from the firing position, and dancing perhaps, women perhaps . . . If you weren't allowed to shoot in this godforsaken area, you at least wanted to get to the bottom of the secrets.

Sometimes, infantrymen walked past the position and talked about what was happening in Lens. Yes, there were heaps of civilians, and pubs and shops and a cinema, just like at home.

And women too?

Yes, of course, women too, young and old. They walked around freely on all the streets, sat outside the front doors in the evenings, and served in the pubs. In some places, there would be a real shindig with dancing until the morning. And you could have a good chat with them as they spoke reasonable German.

"Look at this," said an infantryman, a fellow countryman of Georgi's, who once ate dinner at the First Gun, "you're amazed, aren't you?" He pulled a piece of paper out of his breast pocket and unfolded it. "I tore this off a house ... Well, that's how officers live. But there's enough for us, of course. There's a brothel behind the church, and if you go into the side street by the cinema, there's another one."

The poster passed from hand to hand.

Price List

A. Drinks:

Sparkling wine Henkell Trocken à bottle	18.-	Marks
Bordeaux Château Lafille	6.-	,,
Hungarian wine	8.-	,,
Beer, one large bottle	1.50	,,
Coffee, cup	1.-	,,
,, a small jug, 6 cups	6.-	,,
,, one large jug, 12 cups	12.-	,,
Tea, one glass	0.60	,,
Seltzer, small bottle	0.30	,,

B. Intercourse

30 marks for the whole night

for 2 to 3 hours in the evening and night	20.-	,,
for 1 hour	10,-	,,
for any hour from 9 o'clock in the morning until 6 p.m.	10,-	,,

Vice Squad, a. B. Treller, First Lieutenant and Adjutant

The infantryman was proud of his message. He put his treasure back in his pocket. "With our wages, such a thing is out of the question, but whoever understands . . ."

Georgi narrowed his eyes: "How much does an hour cost for us, I mean for common soldiers?"

"I say, yes, who understands it. Look, it's like this: the women naturally have nothing to eat. If you give them a loaf of bread, for example, or let's say half a loaf . . . I bet you they'll all spread their legs for it. Ha!"

For half a loaf? Half a loaf isn't much . . . you can easily put that back without starving yourself. Hmm, for half a loaf.

Georgi went to the sergeant and asked for leave to get something in Lens.

The sergeant looked at him like he was crazy. He kicked him out; there would be no leave.

Well, that wasn't the end of the matter. Aufricht had to draw his naked girls even more than before. He couldn't save himself from orders and could easily raise the prices to eight cigarettes.

5

Telegram to General Scheidemann in Warsaw from the Staff of the Commander-In-Chief of the Southwest Front:

The day before yesterday, during my presence in Warsaw, I saw on the streets of the city an unusually large number of officers, military doctors, and military officials promenading mainly with women. This proves the inactivity of these military personnel, their complete lack of sense of duty and lack of supervision on the part of their superiors, who allow such a separation from service. This impropriety must cease starting tomorrow and all officers must immediately return to their unit, where they must remain at all times. You must not forget that we are now at war. By tomorrow at the latest, officers without assignment are to be placed at the disposal of the commander of my Staff so that they can be sent to the units that need replacements. During the war, all officers, and military officials must train their men or perform their other duties. The free hours of recreation are to be spent with the troop units. All excesses must be avoided so as not to set a bad example to the troops and undermine their confidence.

Ivanov
F. d. R.: Senior Adjutant Staff, Captain Sulkowski.

(*Vossische Zeitung*, February 5, 1915)

6

One evening, Master Sergeant Burghardt (who had been assigned to the battery as a replacement officer in Douai) invited all the non-commissioned officers to a celebration at his quarters. He lived in a house all to himself on the right wing of the position. There, in a living room with plush red furniture and a piano, the guests gathered. There was a small barrel of beer. The mood quickly grew.

The men wanted to sleep, but noise and laughter roared from every corner.

Reisiger had a toothache. He had gone to bed particularly early. Now, he tossed and turned because he couldn't fall asleep. He was desperate.

Finally, he tied a handkerchief around his head and managed to calm the pain somewhat.

He was half asleep. Then he heard his name. On the street outside the house, someone called out to him: "Reisiger, come to the sergeant!"

Bloody hell, that was the only thing missing!

"Just don't go there. You can call in sick," whispered Rabs, who had just taken off his clothes.

"I'm probably supposed to be on duty. I can't just shirk it," replied Reisiger. "It stinks." He pulled on his boots with difficulty and buttoned his skirt. He went out into the street.

"Man, make it quick. The sergeant is already getting impatient," the guard shoved him in front of him.

He opened the door to the sergeant's house and pulled his cap off his head.

Burghardt staggered toward him. "You can stand comfortably. You're my guest, aren't you? You're a university-studied war volunteer, aren't you? Well, then you can play the piano too, can't you . . . oh man, don't blather on. Have a beer here, won't you?"

Toothache, falling asleep with difficulty, waking up roughly, drinking awful beer on an empty stomach, surrounded by a horde of slightly drunk sergeants, and then playing the piano: a bitter surprise.

Reisiger didn't know what to say. He turned the empty glass over in his hands.

But what could he do? He sat down at the piano. The keys were sticky and dirty. They also had to drink beer. And all the strings were out of tune.

Reisiger believed that he couldn't do his job better and more correctly in a company of sergeants than by playing "Heil Dir im Siegerkranz."[7] He could only do it with two or, at most, three fingers, but he dared.

The response was a roar of laughter. Burghardt hit him on the head with his hand. "You're completely crazy, man. Why don't you play something funny?"

"I don't know anything, Herr Wachtmeister."

"You know nothing?" Burghardt faltered. "Don't you know . . ." He roared a few notes, then continued, "about the crown in the Rhein? What? You have to play that, got it!"

Reisiger: "Then Herr Wachtmeister will have to sing it again . . . perhaps I can accompany you by ear."

The laughter of the company grew even louder.

"Stolzenfels," Burghardt slurred, "Gee, can you sing Stolzenfels? I mean 'Stolzenfels am Rhein,'[8] I mean . . . you're not a real war volunteer!" But he no longer said that in a good-natured tone, it almost sounded like an explosion. "What do you actually want, what, you don't want to shoot, you don't want war, what are you here for, you

---

[7] Heil Dir im Siegerkranz. The imperial anthem of the German Empire from 1871 to 1918, and previously the royal anthem of Prussia from 1795 to 1918. Melody derived from the British anthem "God Save the King." The anthem never became popular in Germany. "Die Wacht am Rhein" (*The Watch on the Rhine*) was the unofficial second national anthem.

[8] Stolzenfels am Rhein. Sentimental song about the Stolzenfels fortress near Koblenz above the Rhine valley. Successful hit in 1906 and still popular during the First World War.

war volunteers?" Reisiger sweated and looked at the keys. *"Stolzenfels am Rhein?" Am I just playing something? Maybe nobody will even hear what's right or wrong.*

Then, the corona of guests roars. They'd got themselves together and were swaying and singing "Stolzenfels." Reisiger tried the accompaniment. Of course, he did. He had heard the song many times before. So off they went. With the courage of desperation, he hit the keys, wrongly but loudly.

When the first verse was over, the sergeant slapped him on the cheek: "You're a fine boy, you're a fine boy. I mean, if you keep playing like that, I think we'll become great friends, understand? Now, the second verse."

Reisiger played the second verse, followed by the third. When the song ended with the fourth verse, Burghardt's voice became uncomfortable again. He slid Reisiger another glass of beer: "Now drink this up and then we'll play the song again."

While Reisiger struggled to empty the glass, the others were already singing. So, he had to hurry. Oh, it went quite well. And even if the singers finished a few bars before him, they all liked it so much that they immediately started again.

When Reisiger had played the four-verse song of "Stolzenfels am Rhein" for the seventh time without interruption, something strange happened. The first notes of the eighth beginning got stuck in the throat of the sergeant and his non-commissioned officers. Cut off. Burghardt set his glass down on the table so sharply that it shattered. He wiped his red face with his wet hand and said in amazement: "Golly, the hogs are shooting!"

Everyone stood up. There was silence in the room. A familiar sound could be heard in the distance: a sharp whistle. The whistling turned into a shrill roar. Everyone suddenly bent their knees a little. Then there was a crash. The sergeant dashed out of the room. Everyone rushed after him. They stared through the open front door into the street. There, opposite the house, perhaps sixty meters away, a yellowish cloud rose up and then stood threateningly against the night sky.

The sergeant shouted at his guests: "Get to your guns at last!"

Reisiger ran to his quarters. Everything there was asleep. He lit a match: "Hello! Go! Battery to the guns! The enemy is firing!" And louder out of the window: "Battery to the guns!"

Seconds later, the guns were ready to fire. Most of the men were standing in their pants. Burghardt was heard to say, "Tell the telephone operator to ask the observation post what's going on?"

Answer: "There is complete silence in the trench. Should the captain be woken up?"

Burghardt, furious: "The telephone operator should kiss my ass!"

They waited for the next shot. Nothing happened.

Gradually, they froze in their underpants. They made jokes about the alarm. Yes, when the sergeant is drinking. Maybe the Frenchman heard the noise.

After ten minutes, the sergeant roared like an irritated bull: "Battery dismissed!"

7

The telephone network of the German troops in Lens is exemplary. There are fireproof centers in all the suburbs, through which important points along the entire front can be reached immediately via various wires. Throughout the city, there are carefully constructed cable lines underground, secured with railway sleepers or layers of stone paving.

The maintenance and expansion of this cable network are all the more important as the construction of the infantry positions has changed fundamentally. At the beginning of 1915, the trench was still a single straight line, connected to the rear area by approaches. Now, the infantry lives in a grid. Four or five manned trenches run side by side. Built-in zigzags, connected to each other by partially covered passages, moved closer to the enemy by forward trenches.

A failure of the telephone line, which meant inconvenient disruption on calm days, could easily lead to confusion and errors in turbulent times.

In order to keep the telephone network in perfect order, large telephone teams were now needed in all units. They manned the control centers. They manned the terminal stations. They went from telephone to telephone to ensure that every fault was rectified as quickly as possible.

As each telephone set, the lifeblood of the troops was located in specially protected shelters, the telephone teams had a great advantage over their comrades. They are more protected against accidental hits and the weather than the others.

But since the lines are destroyed during fire—and only during fire—every telephone unit is faced with the task of having to patch up when all other soldiers are ordered to take full cover.

F.A.R. 96 formed several centralized squads from the crews of its six batteries, which were distributed along three points of the front section.

The squad in front of Loos included Rabs and Reisiger from 1/96. The duty: eight hours telephone shelter, eight hours trench, eight hours battery position, rest.

8

Reisiger preferred to go into the trenches. As an artilleryman, you know very little about infantry life. That's how he got to know it. He realized that it was worse than that of the artillery. Especially worse when he drew comparisons between the accommodation his battery now had, for example, and the miserable holes they had to live in in the trenches. And it was almost a satisfaction to him that he lived no differently from the infantrymen here at the front.

The front line lay a few hundred meters from the last houses of the Lens suburb of Liévin, nestling in a meadow. Toward Loretto, it had to cross a coal heap that had burrowed its black ridge deep into this plain. It was impossible to extend the position on the slag heap because of the coal rubble; palisades made of tree trunks had been erected, behind which only the most essential posts were posted during the day. The rest of the troops stayed in the tunnel that had been driven through the massif.

The artillery telephonist was also stationed here by day in a bulging, braced hole. They squatted in the straw, with the telephone next to them on a box. His only task was to call for fire at the request of the infantry if necessary. It almost never happened. They usually slept until dusk.

At night, a group of infantrymen would take up a post behind the palisade at the top of the coal hill.

They would sit there, leaning against the wall, talking to their neighbors, watching the flares chasing each other in the foothills, dozing off. There's hardly any danger here; the enemy can't get up here without you realizing it early. Certainly, his saps have crept up to the nose of the slag heap, but whoever wants to advance here already has an opponent in the rubble who will throw him back again and again.

Reisiger was up here for many days.

Sometimes, the enemy shelled the trenches. There were infantry casualties and wounded. "C'est la guerre." They shrugged their shoulders. "It's not that bad." If you compare the tremendous fire that was constantly raining down on the neighboring section from Loretto with the short raids on the trenches here, it's fortunate to be here; it's a good, quiet position!

One night, Reisiger's duty was such that he took up his post on the slag heap at midnight and was to leave it again at eight in the morning. The hours in the dark were normal. Rifle shots from time to time. Occasionally, hand grenades from above and below between the trenches. Around four o'clock in the morning—it was just getting light—Reisiger packed up his things: the blanket and the sandbag with the remaining rations.

Then, the enemy began to fire heavily with artillery. You could see that he was firing into the trenches. The shots were spread over a width of two kilometers to the right of the slag heap, some of them hitting the target immediately, others not too far before. You could see how these misses were systematically corrected.

A short time passed before the trench was smoking under the enemy batteries' target fire. The infantry became nervous—came to life. The crews crawled out of their holes hunched over, and the squad and platoon leaders scurried back and forth.

The enemy fire became more intense. They could no longer believe it was a coincidence. A command went up to the slag heap: "Prepare for battle! The enemy is attacking."

Reisiger called the artillery center and was put in touch with firing position 1/96. He learned that the battery had been on alert for half an hour on regimental orders. The constable spoke to him, "You stay where you are. If the infantry demands fire from us, report it immediately. If the telephone line is destroyed, shoot three red flares! But make sure you stay at the top of your post."

And he hung up. Reisiger knew what to do.

No shots had yet been fired at the slag heap. Reisiger sat as if in a box, he could watch what was happening undisturbed.

The enemy fire intensified. The shots drummed on the trench. The impacts came together in an incessant thunder. The lime flew up, boards and tree trunks were hurled against the sky.

Next to Reisiger stood five infantrymen, motionless, rifles at the ready. They suddenly moved together. A group appeared from the tunnel. A machine gun was brought into position. The operator squeezed his way between the trunks of the barricade, scooped up black dirt to the rear, and pushed the machine gun into the hollow.

The haste of the infantrymen and the pallor of their sweating faces were arousing. Nobody spoke. They acted with their mouths clenched in ghostly silence.

Suddenly, the roaring smoke roller was lifted from the trench. The enemy pulled it back almost to his own position.

At the same moment, a sergeant leaped at Reisiger and shouted, "The enemy is attacking. Artillery rapid fire!"

Attacking? Reisiger automatically picked up the phone and repeated the sergeant's words. As he hung up the receiver, the thuds of his own batteries were already hacking into the enemy's roll of smoke like biting dogs.

Was it too late?

The smoke roller began to move, advancing toward the German trench again.

And every time it jumped up, Reisiger saw people running behind it. *The enemy! That's the enemy!*

He pushed the M.G. crew: "There ... Frenchmen!"

The answer, grimly: "Let them come ..."

Reisiger swayed with excitement. *They are coming!* There, the cloud of smoke, closer and closer, and behind it always these running, jumping, gesticulating people. "Shoot, for God's sake!"

The M.G. leader turned to Reisiger. "You haven't been through much yet. Why don't you let them come?" Then he looked ahead again. The cloud of smoke was moving. Another meter closer. And behind it, soldiers.

And then, suddenly, as if a storm had blown them away, the cloud was no longer visible.

And then the incomprehensible happened. As the last of the smoke cleared from the ground, the enemy stood and lay and knelt and crawled and ran and jumped, a gray, living mass. And charges, hand grenades raised, bayonets at the ready, toward the trench.

The machine gun next to Reisiger yelps. Rapid fire from all the guns crackles beside him.

God, what's happening! Dozens of Frenchmen throw up their arms and fall backward to the ground. But other dozens, tightly clenched, continue to press forward.

Hand grenade fire hisses. The flames of the artillery race. And Frenchmen, again and again. Frenchmen: forward.

At the machine gun, they shout in confusion. Reisiger doesn't understand a word. Sometimes, the soldiers laugh, and the machine gun commander points out a new target, and one of them triumphs, "The gang of scavengers won't get up again."

Yes, yes, the French are already in the trench!

Reisiger sees five of them jump over the breastwork just before the tunnel entrance.

One of the Germans down there raises his arm with his hand grenade cocked. Reisiger sees the bayonet of a Frenchman go into his neck. The hand grenade explodes. Both fly into the air, shredded.

A French officer is running around a breastwork. The wide-open eyes! The wide-open mouth! A German lunges at him. The officer raises the butt of his rifle. Before he can strike, the German grabs a short spade and strikes. The Frenchman rolls backward with his head split open.

German artillery fire dances between the trenches. But there is no more enemy there. Only the dead are killed there for the second time, hurled into the air, crushed.

And what else is in the trench?

The machine gun crew next to Reisiger abandons their weapons with a howl and chases down the slope. The sergeant already has the neck of a Frenchman between his hands. He hits his head against the wall and throws him down. Then, they search the position. There are no more living Frenchmen to be found.

The attack is defeated.

When Reisiger was relieved, he heard that the company in front of his section had suffered few casualties. "Only eleven dead."

These dead had now been placed in the approach trench. He had to pass them.

He was hungry, yes. He was allowed to go to the battery now, to rest for eight hours, yes. Everything was certainly in order; he was alive, and he had even experienced something, yes. When he gets to the position now, he might be greeted with a certain amount of respect, yes.

He walked past the dead and then walked very slowly. He thought that tonight, the army report in Germany would say that an enemy attack had been repulsed with great losses for the enemy and that our losses were low. Of course, eleven men don't matter at all. We have an army of millions. It is very understandable that one speaks of small losses.

But he had looked at the first of those eleven men. He was an older soldier with a full beard and a wedding ring on his right hand.

Reisiger did not understand.

9

Now is the time, like-minded comrades! The great but terrible time is making even the worst doubter, the most hardened materialist, realize that all the heavy sacrifices the German people are now making would be in vain if everything were over with death. Therefore, fellow believers! Be active in word and deed for the high doctrine of life after death, the spiritual doctrine for which we have been fighting a peaceful battle for decades. The undersigned publisher is happy to provide any desired quantity of sample issues of "Psychische Studien," leaflets and publishing directories, and is also prepared to mail them out. The ground is prepared everywhere, usually only the seed is missing!

(Oswald Mutze, publisher, Leipzig-Lindenstraße 4)

10

Life in the firing position went on as usual. Even more often, mostly at night, fire was requested from the artillery observer in the trench. But people got used to this very quickly. Distances and aiming points in the trench network were defined on maps for all the guns. When it was set in motion, the machine now had prescribed routes. Every gunner knew what to do at every command, even in his sleep. Very unexciting. It was thought that fate must have been particularly kind to 1/96. Their section of the front was the only one where peace prevailed.

One lived a good day. There was now a permanent battery officer, a young lieutenant, Steuwer, who lived in a room in the house next to the sergeant. He also disturbed the peace very little.

The captain himself was rarely seen. He clearly showed his dislike of the position. It was too boring for him. He didn't enjoy it.

He lived with Fricke, the second lieutenant who had been assigned to him after Busse's death, in a house near the entrance to the infantry position, a good twenty minutes from the battery. That was a quirk of his.

He had located the only building that was still intact, a white cottage, low and unpretentious, surrounded by ruins. The front to the enemy was secured by four layers of upright dug-in trees.

Little was known of Fricke. He had once been in the battery. Apart from his corpulence, nothing was noticeable about him. But rumor had it that he was a dashing dog and a perfect match for Mosel. The two of them often prowled among the shot-up houses in broad daylight, never realizing that the streets of Loretto were in full view of the enemy.

11

The 1/96 Protzen were stationed in the village of Annay, about an hour's walk from Lens. The drivers who brought rations and mail to the position in the evening gave enthusiastic accounts of their new home. It was like being at home, they said; life couldn't be better, and the war could go on for another ten years from their point of view. And then you must drive through Lens every day when you're posted up here. And there, hah, all sorts of nice things!

That's what the crews in the firing position were thinking. The desire to go on holiday to Lens became more and more insistent. Or: Quiet in Annay—there's nothing going on here after all.

You could see with your own eyes that Lens was paved with heavy artillery at times. But if you're unlucky, that can happen here too. And it's no reason at all to satisfy your longings.

One morning, Reisiger came out of the trench feeling depressed. The night had been unpleasant. The infantry had suffered losses. Some of the wounded had been placed in the artillery dugout. They had died there next to him. In agony, he made his way to the firing position.

At the captain's house, he was called by the lad. Mosel himself appeared. *Where was he going? To the firing position, relieved from the observation post in the trench. Would he like to do Mosel a favor?*

To order. There was a shop selling fresh peaches near the church in Lens. He should take the lad's bike, ride there and pick up a dozen.

Lens, Lens! There was no thinking about it, despite tiredness and depression. Besides, a request for a private favor is nothing more than an order.

On the bike, and off we go.

Ah, for the first time in a real city in a year. Wide cobbled streets, beautiful white villas right at the start, aha, the princes of the coal mines once lived here. Large gardens, beautiful plane trees. All of this is not well kept, but there is still a lot of peace and quiet. Sometimes, of course, a wall has been knocked out, and there is a huge hole in a roof with burnt ribs staring out of it. But you can't see what you don't want to see. There is something like peace here. A stroll: lots of military, infantry, sappers, hussars. Lots of cars, mostly with senior officers.

And above all: civilians.

Reisiger takes great pleasure in ringing his bicycle bell at every suitable opportunity. That, too, is peace, just like at home. This stupid woman with her child should pull away from the embankment when a cyclist comes along. The old man might as well put his dog on a leash.

Reisiger pedals faster and faster. It's a pleasure to be a cyclist. Of course, you're a soldier, but on a bike, you don't put your right hand to your headgear when a superior comes along. Instead, you just lift your head up a little and look at him, all civil.

You look at him beaming and happy, a soldier, but now a cyclist in a city with a clean, wide street and surrounded by real, lively civilians.

These civilians are, there's nothing to be said against it, home. The old woman with the pram full of firewood is home. The two children playing with a roundabout are at home. The lady over there with the elegant dress, certainly a lady, is home. The young girl next to her is probably her daughter. They're both laughing; if you don't ring the bell now, you're sure to bump into them.

Round the corner of the house: the main street of Lens. There's the cathedral. And to the right and left, civilians and military, laughing and chatting. And lots of shop windows! There's everything in them. There's cheese, there's smoked ham over there, Reisiger drives slowly, stacked up in a pyramid. Here on one side a sign: Field Bookshop, there a café, there a hairdresser's, the golden sign Coiffeur, supplemented by a poster painted red, Shaving and Haircutting.

Reisiger doesn't see why he should drive through all this splendor. He dismounts and walks on the pavement between civilians and the military, pushing his bike in the gutter.

Aha, the cinema. SOLDIERS' CINEMA written in large letters across the building, with magnificent posters underneath, three comedies with Mäxchen Lindner, admission for sergeants and corporals thirty pfennigs, for privates fifteen. Playtime from eleven in the morning until nine in the evening.

Reisiger is bursting with envy; there is a crowd at the box office.

Next, the officers' mess, the local commandant's office, a dozen military offices, a railway administration, a pioneer park administration, a colliery administration. And now, a particularly large window: the fruit shop.

Reisiger places his bike at the curb and enters.

The shop assistant speaks German, that is, she labors, and where she can't find any vocabulary, she makes up for the lack of it by making impetuous movements with her hands or, if that doesn't help either, by laughing brightly. Absolute peace. A lovely girl. Not only should you buy peaches from her, but you should also take a pound of apples with you and maybe some blue grapes here, oh, and a tin or two of pineapple back there. And for goodness' sake, there's sausage and real butter too.

Reisiger has money. He thinks for a moment. *Oh, buying takes time. Touching and smelling every single apple.*

And then he walks up and down the store with loud steps, his hands in his trouser pockets. "Oh, Fräulein, now pack another knob of butter, or let's say two, and then maybe half a pound of grapes, or let's say a pound. Yes, and of course, a tin of oil sardines ... that's enough for me ... yes, and now pay, payer, *s'il vous plaît.*"

Paid, with a sweet smile from the sweet girl, the peaches for the captain, and his own. What a pity to have to leave here. "Bonjour, mademoiselle." He laughs, and she laughs.

Now, he gets on his bike and pedals off hard—

As he is about to turn the corner again, into the street that leads to the captain's flat, two field gendarmes appear. One waves his hand, instructing him to dismount. "Don't be in such a hurry, my boy, the street is under fire."

There's a crash, and Reisiger sees smoke coming out of a house at the back.

"Is that often the case? I didn't hear anything."

One of the gendarmes spits to the side. "They have their tantrum every day. It never lasts very long, but they usually plough off about twenty or thirty shots."

The other gendarme: "We couldn't care less. They only ever shoot their own countrymen dead."

The first gendarme continues, "Don't swear, Kamerad, don't swear. I'll just be surprised if they put a few things on the main road for a start. We'll live to see this nonsense . . . that's what I say."

Reisiger sees a shot hit the road a few more times. After a quarter of an hour, it's quiet.

He gets back on his bike and rides off.

Strange madness, he thinks. *Strange madness.* On one street, the artillery is pounding into people and houses, and on the next, there is life and hustle and bustle as in the most peaceful times. Two zones five minutes apart, death and life. And one knows nothing of the other or doesn't want to know. *Madness, for Christ's sake, madness this war.*

Just now, he drives past a house whose roof is glowing. Black smoke boils above it. The shot has been well placed and has broken through to the cellar. Whoever was in the house will be pulp.

But civilians walk past, don't look.

And there are the two children playing with the roundabout again.

A strange, horrible madness.

Reisiger brings Mosel the peaches, then takes the errands he had put down earlier and goes to the firing position.

All the loot is divided up there.

In the evening, Reisiger thinks of the shop assistant. A blonde girl with red lips. The dress, of course, was no longer beautiful; it was gray and torn. But the blouse was white. And when she laughed, her breasts moved. Soft.

He asks Aufricht for a drawing. "For two oil sardines. . . . For four, I'll draw a naked one . . . you finally acquired a taste for it in Lens?"

Laughter roars between them.

Reisiger pays with four oil sardines.

Chapter VI

1

One morning in September 1915, the regimental doctor appears at the firing position. One senses that one is to be immunized. Damn it all, you'll have a fever again after the carbolic stallion has driven his rusty needles into your chest, and lousy muscle aches as well.

But how can you avoid it? Lieutenant Steuwer goes around the front with the sergeant and uses a list to check that everyone is staying put. Even the guards have to assemble.

A medical sergeant comes with the doctor. He has two bags with him, from which he takes long, metal boxes painted gray. Everyone gets one of these boxes. When the distribution is finished, the lieutenant commands: "Attention," and the senior doctor gives a speech.

"Folks, you're getting gas protection today. We're going to practice how to put on and use this protection. But first I want to explain to you what it is, if you're interested. You have read in the army reports that the enemy has been unleashing harmful gases on us for some time. He hopes to confuse us with the gas clouds. I would like to make it clear that the so-called gases are just a ridiculous thing that cannot harm anyone. Your eyes water or you get a cough when you get into such gas clouds. That's all. The enemy has, of course, achieved nothing with this nonsense so far. But he has forced us to use gas now. Ours is not quite so harmless. Of course, we abide by the laws of international conventions, which are often trampled upon by those bastards over there, but we will make hell as hot as possible for them. They can rely on that.

"Our gas is blown off when the wind is in the enemy's direction. If that doesn't work, if the wind changes direction, then we'll put on our gas shield on such occasions in the future so that we don't get our own clouds in our faces.

"Our gas protection consists of a gauze bandage that we soak in a liquid. We put this gauze bandage in front of our mouth, preferably so that our nostrils are also covered. Then nothing can happen to us ... I'll show you now."

The doctor opened one of the gray lacquered boxes and took out a bandage. It looked like a compress that students in fraternities put on when they are wounded in a saber fight. A thick gauze pad with a white cord on either side.

He took a small bottle from the tin box, uncorked it, and poured some of the liquid onto the gauze pad. "You have to moisten the bandage so that the liquid almost oozes out. Then squeeze it out lightly and put it on ... Do that once."

That was something new. The battery opened the boxes, moistened the bandages, and placed them over their mouths as instructed so that the nostrils were covered. Most of them had the liquid swimming down their chins. It had a nasty odor.

Then they looked at each other. Very funny. The gunners who had mustaches looked the most mannerly. You could think of mustaches. But the clean-shaven ones looked stupid like at a masquerade party. Everyone's eyes were grinning. Finally, one lifted the bandage and spat. Yikes, that liquid isn't exactly the right thing for thirst.

And now we practiced on command.

"Gas protection off." A laborious search for the painstakingly knotted strap began. The execution of the command did not work.

"Pack gas protection." They rolled up the bandage and put it in the tin box.

"That must be much quicker. So now, once again, gas alarm! Battery, put on gas protection! Now take the bandage with the cords in both hands and, shoo, shoo, tie it like lightning in front of your snout ... one ... two!"

It went better. The colonel repeated his command a dozen times. He seemed very proud of the power of his voice. Then he turned to Steuwer: "It is necessary, Herr Leutnant, that the men have a gas roll call at least twice a week!" He reared up like a cock about to crow: "Listen up, everyone! Anyone who loses their gas protection or doesn't always have it with them will be mercilessly locked up if I catch them. Apart from the fact that it can very easily happen that he pays for his recklessness with death. And one more thing: instead of the liquid, you can also soak the bandage with urine. So, if you've poured out your bottle, you'll have a good soak on the gauze. Understood. Thank you very much, Herr Leutnant."

"Stand still. Dismissed!"

When the gunners were back in their quarters, the new wonder was marveled at and discussed at length. Gas? Of course, they had heard of it. A big mess! That's all that was missing! Eventually, they'll get the idea of breeding germs in bottles and having them dropped by planes. This war is already a god-damned nasty thing!

But when, out of fun and curiosity, the gas protection was strapped on again, it was actually just amusing and funny. And they tried out how to dress up as silly as possible with the help of the gauze bandage. Some put it around their foreheads, others on their noses, and most of them played at being students. There were some at each gun who thought the invention was so new and appealing that they had their comrades photograph them as quickly as possible so that they could vividly convey this marvelous nonsense to their loved ones back home.

By the way, gas attack? There had already been army reports about it in the spring, and infantrymen who had been there had been spoken to afterward. Much ado about nothing! All nonsense! If the enemy doesn't come up with anything better, he can just stay at home. He isn't impressed with such trifles. A real grenade is worse than a stinking cloud. Now we also have gas protection. What could happen to us?

2

March 1, 1915 *Grand Headquarters*

Western Theater of War

At one point on our front, the French again, as a few months ago, used shells which produce foul-smelling and suffocating gasses when detonated; no damage was done.

April 16, 1915 *Grand Headquarters*

Western Theater of War

The use of bombs with asphyxiating gasses . . . by the French increases.

April 17, 1915 *Grand Headquarters*

Western Theater of War

Yesterday, the British also used shells and bombs with a suffocating gas effect east of Ypres.

We receive a letter from the Grand Headquarters:

In a publication dated May 21, the British Army Command complained that "contrary to all the laws of civilized warfare," the Germans had used shells which produce asphyxiating gasses when they burst during the recapture of Height 60 south-east of Ypres. As can be seen from the German public announcements, our opponents have been using this means of war for many months. They are evidently of the opinion that what they are allowed to do cannot be conceded to us. We understand such an opinion, which has no novelty in this war, especially since the development of German chemical science naturally allows us to use much more effective means than our enemies, but we cannot share it. Moreover, the invocation of the laws of warfare does not apply. The German troops do not fire "projectiles whose sole purpose is to disperse asphyxiating or poisonous gasses." (Declaration in The Hague of July 29, 1899.) And the gasses that develop when the German shells burst, although they are perceived as much more unpleasant than the gasses of the usual French, Russian, and English artillery shells, are not as dangerous as the latter. The smoke generators we use in close combat are also in no way in conflict with the "laws of warfare." They do nothing more than potentiate the effect that can be achieved by lighting a bundle of straw or wood. Since the smoke produced is clearly perceptible even in the dark of night, it is up to everyone to evade its effects in good time.

(*Vossische Zeitung*, April 23, 1915)

3

The gasses produced by the bursting of German shells, although much more unpleasant than the gasses of ordinary artillery shells, are not as dangerous as the latter . . .

"The gas cloud hit a section of the front occupied mainly by a French colonial division between Bixschoote and Langemark, causing terror and confusion in their ranks and resulting in a total of 15,000 gas poisoning victims, 5,000 of them dead . . ."

(Hauslian, Der Chemische Krieg, Mittler u. Sohn, Bln. II. edition page 12)

Since the smoke produced is clearly perceptible even in the dark of night, it is up to everyone to avoid its effects in good time . . .

"Caused a total of 15,000 gas poisoning victims, including 5,000 deaths . . ."

4

Let us not forget the unscrupulous glee with which the enemy and the American press announced last autumn the great French inventions that would make it possible to increase the destructive power of artillery shells by the use of poisonous gases. And think of the notorious advertisement of the Cleveland Automatic Co. in which the German translation of a new shell literally reads:

"The material is of a very special kind, of high ductility and strength, and has the property of bursting into small pieces when the grenade explodes. The ignition setting of this grenade is like that of a shrapnel, but it differs in that two explosive acids are used to detonate the charge in the cavity of the projectile. The combination of these two acids causes a terrible explosion, which has a greater effect than any previously employed design. Explosive fragments that have come into contact with these acids during the explosion and wounds caused by them mean death with terrible agony within four hours if help is not immediately at hand. From our experience of the conditions prevailing in the trenches, it is impossible to give medical aid to anyone at this time to avoid a fatal outcome. It is essential to cauterize the wound immediately if it is in the body or head or to proceed to amputation if it is in the legs, because there is hardly any antidote to counteract the poisoning. From this, it may be seen that this shell is more powerful than ordinary shrapnel, as the wounds caused by shrapnel bullets and fragments in the flesh are not so dangerous as long as they have no poisonous admixtures which make immediate medical aid necessary."

Here is a worthy object for the world's indignation. How different the phrases would be if the French or English had succeeded in beating us to it by producing effective smoke generators!

(*Vossische Zeitung*, June 23, 1915)

5

There was no sign of gas in the Loos front sector. Reisiger listened in on the infantry every time he was in the trench as an artillery observer. They knew nothing. In any case, the enemy had certainly not dug in any gas cylinders so far; that would certainly have been obvious from the changes in the terrain.

No, no, everything was very normal. No increased artillery activity. No changes to the wire obstacles. Our listening posts could approach at night as far as they wanted: There was peace over there.

The only new thing was a heavy gun that had been firing for days and with a considerably bigger caliber.

There was a dud near the officer's flat. Reisiger always walked past it with shy respect. He was 1.78 meters tall, but the beast lying in the street was a good 10 cm taller. He didn't know its girth. The diameter of this shiny black, sinister creature could be read. On the bottom of the slug, it said: 38.5 cm. That was the diameter.

Thank God there was only one shot every evening. "The evening shot" was what 1/96 had christened it. It came at seven o'clock sharp. Reliably, to the second.

The five minutes before that were bad.

The same scene every evening. You never knew exactly what time it was all day. But strangely enough, the whole battery sensed that it was five minutes to seven. Five minutes to seven was the time when they put the clock on the table at dinner. They went on talking quietly. But the conversations were already getting a bit more ragged. Three minutes to seven, two minutes to seven. Then someone always said, "Now the trolley must come soon." Then the knife was put down, the outstretched legs were pulled up, and the coziness suddenly stopped. They sat in military style. They no longer spoke. Someone drummed on the table; someone ran his hand over his hair, his finger between his neck and collar.

Then it was seven. So much seven that you could set the clock by it. Then!

A muffled shot. No one acknowledged it. Maybe they counted silently. One, two, three . . . six, seven, eight, nine—

Certainly, nobody counted any further.

A swaying went through all the houses. Not imaginary, no, a real swaying. All the walls were shaking. All the walls swayed in a steady rhythm. The floor rose and fell in a steady rhythm. Plates that were not stationary danced on the table. Spoons that weren't fixed buzzed. Windowpanes rattled, doors swung, and sometimes one popped open unexpectedly.

All this in seconds.

The noise that had been slowly rising became a rumble. This is how an empty ten-meter furniture van drives at a walking pace along a narrow, bumpy road. The horses jump senselessly. Senselessly, the bumpy pavement suddenly becomes smooth tarmac, on which the horses tear along at a gallop.

A thunder, a roar!

Then, the watch owners put their watches in their pockets. Color returns to their pale faces. Everyone looks at each other with a broad, ashamed, embarrassed, horrified grin. Then their mouths finish chewing the bread with tinned liver sausage or kippers. And you say: "Oh yes, this war" or "Well, when will the mail finally arrive?"

We didn't really understand what the enemy wanted to achieve with his evening shot. It always had the same target, was aimed with precision, did not scatter, did not increase in distance, and did not break off. The same target night after night: a railway embankment to the left of the firing position, on the road from Lens to Béthune.

But what for? The railway hadn't been running for months; it was impossible to use the line: it ran between the trenches.

Shortly before seven, the gunners always thought that the heavy gun might just end up swiveling toward the firing position today, that the rumbling trolley would have to sweep away three or four houses . . . Nothing happened. The beast kept eating its way into the mangled rubble of the embankment. And the damage was zero.

After all, you could get used to it, as if it had to be that way. Yes, there were some cheeky people who crawled out of the position on their stomachs just before seven to see the impact with their own eyes. And above all, to be on hand if the shot didn't explode.

Because a dud is cash. You take your sidearm and ax, sit astride the fat beast, and rob it of its guide ring.

It provides good material for about a dozen bracelets. It is made of the finest reddish-brown copper; copper is soft and easy to work with. A bracelet fetches thirty to forty pfennigs; indeed, if you are skillful enough to model an iron cross and scratch the year "1914–15" underneath, each piece might sell for a mark.

6

Regimental Order

The regiment's batteries have done their duty in the battles of recent weeks. Especially 2, 3, and 6 F.A.R. 96, which were under heavy enemy fire day and night, have done excellently. I did not expect otherwise. I give this order: In recognition of their performance, the gunners will take turns moving into rest quarters with the Protzen for three days each. Up to six men from each battery may be ordered to rest at the same time. The readiness to fire is, of course, guaranteed under all circumstances. The crews at rest are to be released from all duties.

von Throtha
for the correctness: Linnemann
First Lieutenant and Adjutant

This regimental order was announced to 1/96 with an addition by Captain Mosel: "The 1/96 first replaces the Gunners Rabs and Reisiger with two other telephones (trench observers). The two relieved soldiers immediately move into rest quarters at the Protzen. Report there at two o'clock this afternoon. Accommodation and provisions will be arranged by Sergeant Hollert."

Rabs and Reisiger marched happily to Annay. Hollert was affable. "I can't give you a palace room, but there's a loft room above the writing room, make yourselves comfortable there."

They climbed up a narrow, squeaky staircase. There was a chamber on the upper floor, dirty and dusty, crammed with junk. It looked gloomy. But there was a cannon stove in one corner where you could cook, and that was a joy.

A small door led into another chamber. It had sloping walls but a bright window and two beds, two beds made of brown mahogany wood, stood at one side of the room.

Reisiger squealed with delight. Rabs whistled shrill signals. Attic room? "Jeez, Reisiger, just like at Adlon's[9]. All that's missing is old Herr Adlon himself swiveling the chamber pots."

They threw themselves on the beds so that the springs howled.

What to do?

Hollert had said that there was food in the drivers' quarters of the First Gun. Nevertheless, Reisiger wanted to cook. "I'll make dinner, Rabs, so you can lie down. Imagine jacket potatoes, washed really clean, boiled in really clean water. And maybe fried cheese or stewed potatoes or fried potatoes . . . what? We're going to the village now, and wherever there's a canteen, we'll go shopping."

Rabs was more in favor of washing. "You can use the oven all night for your frying pans as far as I'm concerned, but now it's time to make hot water. I can't even remember how many months ago I last had a bath."

Bathing? That's fine too. But with what and in what? Search the house first!

They took off their boots and crept down the stairs in their socks. There was a sign in the hallway that read: CLERK's OFFICE 1/96, KNOCK!

For God's sake, don't go in there. So, out into the courtyard. *Aha, there was a weathered rain barrel next to the well. Is there any danger? No. So! The barrel was tipped over; brown water splashed over the sand. The bathtub is found.*

---

[9] Adlon. Luxury hotel in Berlin at Pariser Platz/Brandenburg Gate. Opened in 1907, destroyed in World War II, re-opened in 1997.

But how to boil water in it? There was only one option. Back there was the stable for the sergeant's horse. Rabs looked inside. The horse looked around with interest. "I suppose you'll excuse us for borrowing your bucket for a moment," said Rabs, grinning at the horse.

Then he trotted off with the bucket.

Afterward they bathed. While one of them crawled naked into the half-full, unfortunately somewhat leaking barrel, the other had to douse him with the bucket and then scrub the trembling body of his victim with plaited straw.

The roles were reversed. The result was amazing! The water had taken on a deep coffee-brown color. Rabs and Reisiger, on the other hand, looked as delicate and white as young girls.

But what now? They stood next to each other, teeth chattering and dripping. The small towels they had in the bags were not designed for such extensive personal hygiene.

"It's best if I unbutton the curtains, and we wrap ourselves in them," said Reisiger.

"So that the constable gives us notice to leave in the morning," replied Rabs.

"The simplest solution is, of course . . ."

"We're going to bed! Why shouldn't we go to bed? I'm sure there are sophisticated people in Germany, for example, who, if they're staying in a hotel, will go to bed around three in the afternoon if they feel like it."

Rabs went on: "They'll go to bed with a big cigar in their mouths, or they'll have the waiter bring them beer and warm sausages." He slapped Reisiger hard between the shoulder blades with his giant paw, and continued, "I bet, my dear, we'll live like bankers here, and you'll see, tomorrow afternoon, I'll personally get the beer and sausages. That would be a laugh!"

Whinnying, they jumped into their beds. Gosh, soft and with wedge pillows. They wrapped themselves in their blankets. After a few minutes, they were dry and reasonably warm and could put their shirts back on. Reisiger conjured up something to smoke, and then they lay there for hours, not taking the cigar out of their mouths, letting the ash fall quietly and talking languidly, almost paralyzed by well-being.

"The most important thing," Reisiger said, "is that we have a set program for the three days. I'm in favor of sleeping until ten in the morning, then going for a walk, then sleeping, then going for a walk, then sleeping. And always eat in between."

Rabs blew a puff of smoke at him. "The question is whether we're really off duty. We don't know what the constable has in mind. There will be a roll call after all."

Reisiger nodded. "There has to be a roll call so we can get the mail. But nothing more will happen. The drivers can work here. The right thing would be for every gunner to be given a driver as a valet as soon as he comes to rest. I'm already afraid of cleaning boots."

Pause.

After a while, Rabs blew smoke through his nostrils and said, "Hmm."

He was surprised that Reisiger didn't answer.

Then he thought, *It's difficult to answer.* "Hmm." But he was now too tired to speak another long sentence. At last, he decided. "Yes, my son, you should have a servant."

Reisiger was still silent. Rabs was annoyed. *Stupid dog, he's sleeping*. He shouted, "To the guns!"

The next moment, Reisiger threw up the duvet and stood helplessly in the room in his shirt.

Rabs roared with delight. "You're the biggest idiot of this century," he neighed. "Gunner Reisiger reports the battery ready to fire ... Don't let them snatch you away ... You can't sleep here when I'm talking to you."

Reisiger squinted against the bright light of the window. He sucked on the cold stub of his cigar, which still sat between his lips, and grumbled, "Old monkey." Then he curled up again with a loud noise and went back to sleep. Rabs was quickly infected by his snoring and snored too.

Unfortunately, it was almost dark when they were woken by the call "roll call." They dressed angrily; they shouldn't have slept through the first few hours of the rest period.

The mail was a comfort. And the drivers of their lorries, who were clearly delighted with the visit, were a particular source of comfort. The roll call had barely finished when they were both urged by their comrades to tell them about the news from the front. They also learned that the infantry canteen had excellent beer. "Reisiger, you're a war volunteer, how about a round?"

It wasn't possible to avoid it. One round turned into several. Rabs and Reisiger were glad when they finally managed to escape. They stumbled into bed, smoked their last cigarette, and read the letters again. Reisiger even had some newspapers. They were months old. But they were good enough for reading aloud. He was particularly fond of adverts.

"You, Rabs, do you know that we've thoroughly won again? Listen, here, Berliner Tageblatt, Weltspiegel, November 1, 1914:"

German Victory

German victory in the field of women's culture is unstoppable. The renunciation of everything foreign brings it as a natural consequence, and the fortunate fact occurs that the pernicious effect of the French corset fashions, which had made almost all German women sick, has come to an end. The only German product that is not based on French models and that only achieves the development of real beauty without damaging health is the long-known Thalysia Edelformer. Thalysia Edelformer. It is not laced, does not hinder breathing or freedom of movement, is not cumbersome, and is not as sinfully expensive as the Parisian corsets. On the other hand, it is also not as shameless as the Parisian corsets but transforms even an overly voluptuous form into something delicate and German-minded. In other words, the Thalysia Edelformer is a genuinely German, hygienic marvel...
Thalysia Paul Grams, Leipzig

Rabs neighed with pleasure. "Gosh, gosh, I don't know anything about newspapers, but that's quite something. You'd have to go there and drive through it."

Reisiger had lost his good humor. "That sucks. But that's how the home warriors do it. And do you think that's only the case in Germany? Oh no, there is this mixture of patriotism and business in all countries. Do you want to hear more?

Here, this is also quite nice, also a Berlin newspaper, September 14, so that you know how to wash yourself in the future":

Wash Yourself Without Water and Without Soap with Kiri

Indispensable and most beautiful gift of love for our warriors in the field. A little "Kiri" removes even the worst dirt from the face and hands in a minute, and our warriors feel clean, refreshed, and reborn after using it!

"Stop it," grumbled Rabs. "You should put so much Kiri in their mouths that they forget to wash. Jeez, jeez, and that's why we're lying here in the dirt. Well, good night then, I'm off."

7

Reisiger is dreaming:

"We're getting to northern France now. Meier, put up the map!" Student Willi Meier hurries to the lectern and lowers a large special map of France from the map stand.

Head teacher Wittig grabs the long bamboo stick, waves it around in the air, and finally hits various points on the map. He then outlines a specific area with the tip of the stick.

"Reisiger, you're playing again. Get up! What's the name of this area?"

Reisiger jumps up from the bench. He can just hear his neighbor whispering something to him and shouts, "Pas de Calais."

The head teacher hits him lightly on the shoulder with his pointing stick. "Right. Sit down. Département Pas de Calais . . . Cities: Lille, Béthune, Lens, Douai, Arras . . . Repeat, Krüger."

"Département Pas de Calais . . . Lille, Béthune, Lens, Douai, Arras."

Head teacher Wittig bangs between the benches. "That must go much faster; these are names that must become second nature to you. And who knows the places I'm about to show you?"

Reisiger sees that Wittig has the pointer tucked under his arm like a lance. With the tip of this lance, he rides off again and again in strange leaps toward the map. And every time the tip pierces the map at a certain black dot, he shouts a name in an angry voice, "Souchéz! Givenchy! Carency! Ablain! Grenay! Loos!"

The class starts to laugh terribly. Reisiger is also shaken by this laughter. The head teacher's leaps become more and more comical; his voice gets angrier and angrier, and his face is distorted beyond recognition.

Finally, Reisiger can no longer hide his laughter. A few shrill sounds come out of his mouth.

Wittig has heard that. He turns on one heel in a flash, pulls the lance out of the map, swings it up, and hisses at Reisiger, "You laugh, my son, you dare to laugh, my son? I want to teach you these names. I want to teach you these names so that you won't forget them for the rest of your life. That would be something!"

Reisiger breaks out in a sweat. He wants to push his neighbor aside because Wittig is now racing along the class corridor, heading straight for him.

He flees. A few more steps and he will reach the door handle. But then Wittig shouts something. The classroom goes dark, so dark that you can't see anything in the room. And Wittig is already beside him, clawing his hand into his neck and pushing him against the catheter.

The map is lit up snow-white and red lights appearing on the blinding surface, red bubbles that suddenly burst open and shoot out a pointed, yellowish flame. And every time a flame pops up, the head teacher, Flame, presses his fingers into Reisiger's throat and says a name, "Souchéz! Givenchy! Carency! Ablain! Grenay! Loos!"

The hand on Reisiger's neck closes tighter and tighter. The head teacher's mouth is close to his ear. "Will you finally learn that?" he hisses.

"At your command, Herr Hauptmann!" Reisiger snaps his neck back, because he knows that you have to stand at attention in front of a superior. Then he shouts loudly once more, "At your command, Herr Hauptmann!"

Reisiger sits up in bed. His fingers tremble through his hair. He looks around the room. *Thank God. I'm at war; Rabs is lying over there.*

He whispers his name softly.

Rabs rolls around in the straw: "Man, why don't you shut up? You've been talking all night."

Reisiger lies down on his side with a smile.

8

"We have to play reindeer properly," says Rabs the next morning while shaving, "really enjoy ourselves and, above all, don't rush things. Nobody can do anything to us. We have a look around the village and are invited to the drivers for lunch. They serve tinned meat, Hungarian goulash. My dear, Holle can cook; he's a hotel owner or something."

A walk. Nice village, clean, with well-kept streets, small white houses, flower boxes in front of the windows. Including the inevitable geraniums, which are still blooming in September.

Around midday, there is music in the square by the church. The brass band of the infantry regiment has formed a circle, a conductor stands in the center and there is a real concert. First operas, then popular songs.

The place is teeming with military personnel. Everyone has their hands in their trouser pockets, some are sitting on the steps of the church.

When they particularly like a piece—like the overture to "Lohengrin" and the waltz from "Der Liebe Augustin"—they clap.

Officers pass by. Mostly in peacetime litewka, some in high, patent leather boots.

They are barely noticed.

Reisiger and Rabs are strolling through the crowd arm in arm. Reisiger recounts memories of his youth. "Do you know Potsdam?"

Rabs shakes his head. "I know it, but I've never been there."

Reisiger points his hand over to the music. "They had that every day in Potsdam during peacetime. We were students back then. You can imagine how we waited until the last lesson was over at around one o'clock. Then we'd get on our bikes and gallop off. And do you know what the main purpose was?" He slaps Rabs cheerfully on the shoulder, and continues, "The little girls, of course. They came out of school at the same time. And then we met up and had a stroll."

Rabs looks around. After a while, he says, "You, Adolf, there are so many old women here . . . there must be young ones too. Do you see any?"

No, Reisiger didn't see any. There were old mothers squatting in the doorways of the houses. *Where there are mothers,* he thought, *there must also be children. The sons—that's clear, of course, they're over at the front. But the daughters?*

They decided to ask the drivers about the problem over lunch.

The three drivers of their gun had a particularly nice flat. When Reisiger and Rabs entered, their hosts were already sitting at the table.

Greetings. Grinning faces.

Reisiger was very hungry. He waited for one of the three to finally bring the food.

He counted the plates—seven. *What does this mean?*

Then, a woman appeared from the next room with a friendly, embarrassed smile. She was carrying a white soup tureen, nodded a friendly "bonjour" to the guests, and placed the food on the table.

Holle, the supposed hotel owner, stuck his nose deep into the pot. Then he said, "Madame, potatoes, pomme de terre."

The old woman nodded her head vigorously. "Oui, oui, camérad, cartoffln." She laughed at Reisiger and said, "Bon deutsch ich, n'est ce pas?"

Reisiger bowed. "Très bon, très bien, I don't know exactly what it's called myself, madame."

Rabs laughed. "Children, you live here like shit in a flowerpot. We should have a mum like that upstairs. Does she cook everything for you?"

Holle pursed his lips. "Wait and see, the best is yet to come." He shouted after the old woman, who hurried off again, "Madame, and then Marie will come ... My bride, n'est ce pas?"

Madame turned round in the doorway with a grin. "Non, non, monsieur, nix bride, seulement mademoiselle Marie, compris?"

A girl came in.

She had a bowl of potatoes in her hands. She was impartial, approached the table, and nodded slightly to the two gunners.

Holle cleared his throat. "Mademoiselle Marie, this is Monsieur Rabs and the little one here is called Reisiger."

Everyone laughed out loud. The girl was not put off by this. She took a ladle and filled the plates for the five soldiers, for herself and for her mother.

Then they ate.

Reisiger looked at Marie. She had black frizzy hair, parted in the middle, a small face with dark eyes and red cheeks. Her clothes were poor but touchingly neat.

Reisiger thought about it. These people had been living here for almost a year and a half, cut off from their homeland and with no means of providing more than their daily food. Poor people.

He wasn't very talkative at mealtimes. Rabs entertained the drivers all the more.

Holle sometimes reached across the table and tried to stroke Marie's hand when he made a joke. Reisiger saw her mother's eyes shifting restlessly back and forth. Marie always withdrew her hand and smiled. She didn't say a word.

After the meal, the woman brought a large pot of coffee, and Marie put out cups. The soldiers smoked and chatted.

Reisiger couldn't talk. He felt insecure and wanted to leave.

Rabs was annoyed about this. "It's like Mum's here; why don't you stay?" He looked at Holle and said, "Or do we want to meet again in the evening?"

Holle suggested, "If you fancy it, we can go boating. There's a canal behind the village. Well, Reisiger?"

"Gladly."

Holle laughed at Marie. "Isn't that right, mademoiselle, une partie en bateau?"

Marie got up and left the room. "Nix promenade bateau."

9

The boat trip came about. The canal flowed not far behind the village. A beautiful river, delicately embedded in the Flemish gray of a gentle, sad landscape.

The three drivers, Rabs, and Reisiger met in the evening at the railway bridge, which spanned its broken arch over the water at the end of the village. Two old barges were moored on the bank.

Holle took the word with importance. Of course, not all six of them could get into one boat. He suggested this distribution: two drivers and Rabs in one, he and Reisiger in the second. And then they would set off as far as Lens. There was a small pub with real cognac near the military cemetery. "So, meeting point," he called out and gently took Reisiger aside. "Café Imperial Lens. Drive off; I'll depart right after you."

They boarded the boats as he had instructed.

The first ones left and disappeared.

Reisiger was left alone with Holle. Holle hesitated. He barely stroked the water with his oars.

How quiet it all was. A soft wind, a narrow moon in the sky, not yet high, with a faint light. Reisiger looked with delight at the silver crescent.

He was startled. Holle whispered, "So . . . there's a party. I must take my Marie with me, you see, Kamerad. Watch out, she's waiting at the corner."

His Marie? Reisiger said nothing. *His Marie? Oh, the girl from the neighborhood.*

He had a hot feeling in his throat. *So, I'm supposed to go on a boat trip with the girl?*

He heard Holle's voice again. It was remarkably soft, a little hoarse. "You don't have to talk about it right away, Adolf. Look, Kamerad, I'm thirty-four years old. Yes, and at home, well, I've never found the right thing. Well, life is like that, and that's what Marie has become. Don't think about *coucher* and all that. Kamerad, that's not so important. But I really like her. I can tell you that . . . I like her."

Quiet rowing strokes.

Reisiger wants to say something. He doesn't know what.

Holle sweeps the water again.

Then he starts again, "Oh, you mean at the end because of Franzmann . . . ? Oh, Adolf, you can't divide girls into French and German. If you really like them, that's absolutely rubbish, isn't it?"

A very soft whistle can be heard. "There's Marie."

The boat turns.

Reisiger sees her. She stands shyly by a tree. When she realizes that the boat is hitting the riverbank, she straightens up. "Pierre?"

Peter Holle's voice has become even softer. "Marie."

Reisiger presses himself tightly against his seat.

The boat sways. Light kicks. "Oh mon, Pierre . . . cher Pierre."

Marie throws herself willfully into Holles' open arms. He pulls her gently against him.

Reisiger senses that they are kissing.

The boat drifts in the middle of the water.

Whispering, "Mon Pierre"

"Oh, dear Marie . . ."

Reisiger takes the oars. He puts them in deep. The boat takes off with a jerk.

"Well, first say good evening to my mate, bonsoir, Marie," Reisiger hears.

The boat sways.

Reisiger feels a soft hand slip from his arm. Fingers grope. He takes the hand and squeezes it. "Bonsoir."

Holle says, "Guten Abend."

"Gu-tennabt," Marie repeats. Then, her hand unexpectedly reaches into Reisiger's hair. She rummages through it. "Oh . . . gutt Aair, verrie sofd."

Holle laughs. "Watch out, Adolf, she's falling in love with you."

Reisiger puts his hand to his forehead. "But Mademoiselle . . ." he stammers insecure.

He wants to remove his hand. Then she grabs it and holds it tight. She stops.

He senses that these are her breasts. He trembles. The stranger's hand remains on his joint. And he must, must close his fingers.

Holle impatiently says, "Well, Marie, viens ici."

Reisiger senses . . . *It's her warm breasts.*

The hand caresses his joint. The girl presses herself firmly against his body.

Holle: "Marie!!!"

Marie: "Un moment, monsieur." The little voice says this emphatically.

Reisiger feels overwhelmed with happiness.

Nevertheless, he doesn't want to upset Holle. "Why don't you go," he says, begging.

Marie falls onto him, heavy. Her mouth touches his ear, grazes his cheek, presses against his mouth. Sucks herself in.

He realizes that he can't rise, that he doesn't want to fight back. He sets his teeth into the girl's lips. Body burns against body.

The boat sways.

Reisiger regains consciousness.

"Why don't you go and see your Peter," he says harshly and pushes Marie away from him toward Holle.

At the same time, he takes the oars and pulls.

Holle hasn't understood anything. "Well, Marie, sad," he says. "Yes, yes, the war . . ."

Reisiger rows harder.

The girl's voice, almost like a cry: "Olala, c'est la guerre. Grand malheur pour nous et pour vous et pour tout le monde."

Holle gives up the conversation. He takes a harmonica out of his pocket and plays.

Marie has dropped a hand in the water and is dragging it so that the waves hit her up to the elbow.

"Shall we go to Lens?" Reisiger asks.

Holle yawns. "Oh, leave it for today. We're getting off. Marie has to go home too; otherwise, the old lady will complain."

They land. Marie scurries off. "Goodbye, Pierre, au revoir, au revoir, monsieur."

Then, the other boat arrives. Lens was under heavy fire; they were fed up.

Everyone goes back to the village.

Reisiger staggers. That a single encounter can throw me off balance so much, he thinks. *A girl, a girl, a girl—so much to throw me off balance—a girl, kissed me—Marie, I have to—I have to tonight—heavens, what is it, what is war, what is war when a girl has kissed me, a girl—I have kissed . . . I want a girl for once—for once—and there's a war going on and all hell is burning up ahead—if only it wasn't meant for us, because to die now . . .*

And then Rabs and Reisiger lay in the grass along the road in front of the first house of Annay and looked at the twitching sky.

Impossible to speak. They were both thinking the same thing: *Just not us!* They did not say it. They hardly noticed that many villagers had gradually gathered here.

Whispering voices. Now, the fire was coming straight for Lens. Sometimes, you could see black gables standing out sharply against the sky. At one point, it seemed as if the cathedral was on fire. There: the tower, the roof, covered in sparks. And a flame reaching into the clouds.

Then machine guns barked. "They're cooking peas up ahead," said a voice.

Reisiger listened. There is no better comparison; a relentless seething, a wave of sharp, biting blows in a row.

But there was no alarm.

Hollert appeared. "Oh rubbish, it's none of our business!" Then he shouted into the village street, "Forage Master, you make sure the stable is properly guarded. I'll be woken up if anything happens. There's no need to keep the telephone in the office manned. But the battery clerk should strap the headphones around his ears when he goes to sleep."

He disappeared, humming.

The village street gradually emptied. The fire up ahead is not dying down. But it can't be helped ...

Rabs and Reisiger left too.

Rabs fell asleep immediately. Reisiger lay awake. *Marie, a girl, Marie, a girl ...*

What's going on? Isn't there a rumble on the stairs? "Rabs!"

As Reisiger struck a match, the door opened. The battery clerk said, "Orders from the captain: Get into firing position immediately."

They set off on the march.

When they arrived in Lens, the fire attack was over. The town lay in sultry calm.

Only on the Loretto Height had the fire rage incessantly. They saw the glowing massif. They pressed along the houses and crawled through the streets.

Reisiger thought of his dream: *Loretto Height, Souchez, Carency, Givenchy.* He was freezing for a moment. But then he continued to think, *Loos.* And was reassured.

They arrived at the firing position at 2:20 in the morning. Everything was asleep. The guard gave them an order from Lieutenant Fricke to take over the telephone in the battery. Instead of Steuwer, Fricke himself was in the officers' dugout of the position. He was to be woken up if anything suspicious was going on.

The battery's telephone shelter was in the cellar of the house next to the Third Gun. The two telephone operators were asleep with the candle burning. They were happy when Rabs and Reisiger relieved them.

Chapter VII

1

Reisiger urges himself to take over the first telephone watch. He believes that this is the quickest way to dispel the sadness that is in him. Rabs lies down on the straw and falls asleep.

Reisiger makes line tests. "Control center!"

Line in order. They greet each other. War volunteer Brinkmeier on the receiver: "You were lucky you were with the Protzen... there's a lot going on up here."

"It's quiet everywhere, isn't it?"

"In the last two hours, yes. But we got quite a few things here near the headquarters. And the Frenchman shot the trench observation position to pieces."

Reisiger is shocked. *Trench observation? Were there any casualties?*

"No. Sergeant Karl was in front. When the bandits paved over the position, he went to the company commander to get the infantry's wishes. And when he came back, the place was rubble. And everything was gone... coat, rations, telephone." Brinkmeier laughed.

"So, where's the observation post now?"

"In the armored tower, just behind Mosel's house. There's a good cable there, but you'd better not call; the captain himself is in front... Well, goodbye."

Reisiger hangs up the phone. "You, Rabs, our observation post is busted. So, we've got rid of the worry that we'll have to tiptoe back in the morning."

Rabs has no interest in conversation. He grumbles and rolls onto his stomach.

Reisiger leafs through the telephone notebook in front of him. All incoming and outgoing calls must be noted here with exact times. That has become the fashion recently.

He writes: "September 25, 1915" underneath "2:20 a.m., replacement gunners: Rabs and Reisiger."

He turns the page back to catch up on the latest news.

Gosh, there must have been a lot going on. He reads:

"24. 9. 15: 10:5 o'clock in the evening, report to division: Enemy has been holding outskirts under lively fire for several hours. 1/96 returned fire from about nine o'clock in the evening until ten. In the process, a recognized enemy battery at the southern exit of Loos was silenced. sign. Fricke, 1/96."

"Eleven o'clock in the evening. Captain Mosel's telegram to Fricke: Karl has reported that trench observation has been destroyed. He is to send a sergeant to the officers' quarters immediately. Lieutenant Steuwer moves into the observation tower with him."

Reisiger tries to wake Rabs—calls his name several times. No answer.

He continues reading: "11:20 a.m.: Report to Captain M.: Battery has just received 417 grenades and 388 pieces of shrapnel from 2nd L.M.K. sign. Burghardt."

"25. 9. 15: 12:10 o'clock in the morning. To all batteries F.A.R. 96: Infantry has just reported that listening posts in the Loos section have detected heavy movement in the trenches and lively train traffic in the rear. Increased readiness to fire ordered for all batteries, F.A.R. 96."

"12:40 p.m.: Herr Hauptmann tells Herr Leutnant Fricke that he has gone to the observation tower. Herr Leutnant Fricke is to ensure increased vigilance of the guards."

The telephone buzzes. It says: "This is Sergeant Karl . . . Oh, Reisiger, you're here . . . I just wanted to check the connection."

He hangs up. Reisiger thinks he could have said a little more. Then he picks up the book again.

"25. 9. 1:50 in the morning. Report to division: Officer's patrol found out at one a.m. that the enemy had cleared away wire entanglements in front of trenches."

"1:55 a.m. Regimental order: The gunners of all batteries that are at rest are to be ordered into firing position immediately. F.A.R. 96."

"2:20 o'clock. Lieutenant Fricke is to pass on: Protzen of 1/96 are to be alerted immediately but remain with bridled horses in Annay for the time being. Sergeant is to send Captain Mosel's and Lieutenant

Steuwer's horses with lads to the officers' quarters, and also two dispatch riders immediately to the firing position. Mosel."

Reisiger checks his watch. It's just after three o'clock.

He still has an hour before Rabs must relieve him. *What to do?* Something's not quite right. On the other hand, the whole front is quiet now. You can't really believe in an attack. He lights a cigarette and writes a letter home. In between, the phone rings several times. Each time, he thinks he can catch some interesting message, but it is always just a line test. Finally, he calls the armored tower for news. Sergeant Karl whispers that he can't say much . . . the captain has been lying down for a moment, but there's nothing going on.

Four o'clock. Reisiger has become tired. He is cold. He goes to the door and pulls the canvas tighter into the frame in front of the opening. He looks outside for a moment. It is already dawning, and you can see the houses of Lens.

The battery guards chop along the street with loud footsteps.

As he tries to sit down again, the footsteps quicken. What is it?

One of the guards pulls down the canvas and shouts: "Red flares!"

And voices everywhere shout: "Red flares!"

At the same time, the telephone buzzes like mad.

Reisiger picks up the receiver, kicks Rabs with his foot, hears Sergeant Karl, cold and calm: "Battery to the guns!"

He throws the receiver on the table. "Rabs, you pick up the phone." Up the cellar stairs, "Battery to the guns," shouted into the officer's room, up to the second floor, the same command, down the stairs and into the street: "Battery to the guns!"

He is not yet back down in the telephone cellar when Lieutenant Fricke appears. "Report: battery ready to fire!"

A second later, from the tower: "Whole battery barrage!"

Battery fires. The houses shake, lime crackles on the cellar walls.

Rabs: "Herr Hauptmann asks if the battery is taking fire?"

Fricke answers the phone. "I'd like to speak to Herr Hauptmann myself."

Mosel's voice is so loud that Rabs and Reisiger understand every word.

"Do you hear? Attack," says Rabs. They both listen again.

"Heavy fire on the whole front, did you hear that, Rabs? But our battery isn't powdering badly either!"

A man jumps into the cellar from the hallway: "Herr Leutnant, we're taking fire, two shots came in directly in front of the Sixth Gun."

Fricke interrupts the conversation with the captain for a moment. "Battery keeps firing!" Then he continues: "Herr Hauptmann, there are enemy shots into our position."

The captain does not respond. You hear: "Herr Leutnant, battery should fire faster and the distance . . ."

"Fire faster!" yells Fricke. Then he gets nervous. "Hello, hello . . . off . . . Lord God sacrament, isn't your damned machine in order, don't stand around, you stupid monkeys." He slams the receiver down on the table. "I need to reconnect to the observation post."

Reisiger and Rabs examine the device. "Everything is in order. But there's no answer over there. So . . . the line is destroyed."

"Then get yourselves out of here . . ." The whole house shuts down with a muffled crash. "Then get yourselves on line patrol immediately. Without a telephone, we're sitting here, damn it, in a mousetrap. What are we supposed to . . ."

A new explosion lifts the house so that the floor of the cellar tilts. Wood splinters. Pours of shattered windowpanes spill onto the pavement.

Rabs and Reisiger are confused by the lieutenant's rage and by the two impacts. Finally, they strap on the belt and take the inspection device. "Hurry up," says Fricke a little more gently and pushes them toward the road ahead.

At the same moment, a black cloud hits them. A blast of air knocks them against the wall, and they stagger back.

"Damn it all," grinds Fricke. "Then you'll have to try to get past the guns in front."

They all three go up the cellar stairs and step out into the courtyard. The Third Gun is steaming in front of them in the wash house. The roof has been torn off; the tiles lie between the gunners. The gun fires with the precision of a machine.

Fricke goes to Burghardt and shouts in his ear, "I have no connection with the captain."

Burghardt shouts back, "Then the telephone men have to go on line patrol; but I'd leave one down at the phone, maybe the captain will send out a squad himself."

Fricke waves to Rabs and says, "Off!"

Rabs jumps out of the position, crouching in long strides.

Reisiger returns to the cellar alone.

It's scary to be alone. Upstairs you are at least deafened by the noise and the smoke and know who is friend and who is foe. Down here, there is only senseless drumming.

The telephone buzzes.

Why is the phone buzzing? It's broken. He picks up the receiver and reports ridiculously correctly, "This is Third Battery F.A.R. 96."

Captain Mosel, barely recognizable, barely human, shouts over his own voice, "Gas attack!"

2

There is always a fine whistle in the air. And from the depths of the field slope, which descends towards the valley, a cheerful banging sounds continuously as if the shooting galleries of the Munich Oktoberfest were down there.

(Ludwig Ganghofer, "Journey to the German Front 1915")

3

Captain Mosel stands at the scissor scope in the tank tower. He does not know why he has just shouted "gas attack" into the telephone. Subconscious? At the moment he doesn't realize, doesn't understand, and doesn't overlook what has happened and what is at stake. Certainly, the enemy had shelled the trenches. But that was nothing new. And it wasn't even aimed at the infantry; most of the shots were fired toward Lens.

Gas attack? Why?

But . . . but now? Can't you hear a sharp whistling hiss very clearly?

He presses his eyes to the scissor scope. *What on earth is happening?*

Clearly visible, clusters of white clouds are rising directly in front of the first enemy line, one next to the other.

And the hissing? Yes! The hissing is always clear!

The white puffs push into each other above the ground, finally forming a thick, billowing curtain in front of the enemy's first trench.

Can't the infantry see that?

He takes his eyes off the glass and rolls aside the flap of the armored tower above him. He listens.

Always just the hissing. Always just the hissing.

The infantry is not firing.

It is dead silence.

He pulls down the flap of the armored tower again.

"Karl, there's an incomprehensible silence outside . . . take a look through the scissor scope."

And as the sergeant stands up. "If only the infantry wanted to shoot! Bastards!"

Karl has his eyes on the glass. After a while, he turns around, paralyzed, and nods non-militarily: "Herr Hauptmann can believe it . . . a gas attack."

This brings Mosel back to his senses. He pushes Karl aside. The cloud has thickened. It now lies on the ground like a wall. No, it's not lying, it's creeping. It creeps slowly.

"There's no point in our battery still firing into the enemy trench. Call: rapid fire twelve hundred."

Battery! Nobody answers. "The line is shot to pieces again, Herr Hauptmann."

The cloud is getting closer.

What if, for heaven's sake, we could find out what's going on behind the cloud?

"Do we have a connection to the front, at least? It's incomprehensible that the infantry isn't firing, Karl."

The sergeant calls . . . No answer.

What to do? Do what, do what?

Then Karl raises his finger. "Herr Hauptmann, they're firing now."

Yes, the infantry is firing. There is a clatter. It rattles and whistles. Mosel becomes elastic and cheerful. He looks through the scissor scope. "The infantry is firing fiercely!" He sees how the men standing in front of the trench repeatedly pull their rifles to their cheeks; others shoot while kneeling. Still others have lifted themselves up to their navels from the breastworks and are brandishing hand grenades.

All hell breaks loose.

Bravo, great, thinks Mosel, *now things are finally getting the right momentum.*

Only—the cloud? Half aloud: "The only damned mess, Karl, is that you can't shoot the cloud to pieces. It's almost in the trench, five meters at most ... all those damn German batteries are shooting uselessly six hundred meters, or what do I know, behind the gas."

He pulls himself away from the scissor scope with a jerk. "We have to get into the firing position. Come on, let's try to get our horses! Dismantle the scissor scope! Take the phone with you! Go!"

Karl: "But Leutnant Steuwer is still with the infantry ..."

"We can't wait for that. He'll help himself."

As they crawl out of the small opening of the tank tower, they hear the whistling hiss more sharply. Mosel looks over the approach trench and ducks his head. "If we don't reach the battery at a gallop, the gas cloud will be there before we are."

They run after each other. The approach trench is ending. They jump out.

Yesterday, there were six relatively intact houses here. Now, they are a source of fire, with flames leaping out of them.

Go on! If only these hounds hadn't shot the horses!

Past the telephone station.

Mosel is tempted to jump in. There are also some gunners from 1/96 stationed there.

But one look is enough. The heavy concrete cellar is caved in. *So, they're all dead*, he thinks.

Over there, the officers' quarters. It no longer has a roof. The house has been smashed through to the cellar, even though the barricade is still standing.

In one corner of the rubble is a chest of drawers with a green plush blanket. On it lies a picture in a frame: Mosel's wife.

Pointless.

One jump—the house is reached. Thank God, there are lads and horses behind the barricade.

"Herr Hauptmann, we've never really had a fire like this before." Mosel pulls a horse around by the curb, swings into the saddle, and gallops off.

4

The soldier is given the cold iron in his fist, and he should wield it without weakness or softness. The soldier should shoot to kill; he should thrust the bayonet into the enemy's ribs; he should drive the whizzing blade into the enemy; that is his sacred duty, yes, that is his divine service.
("In Gottes Namen durch" by Divisional Pastor Lic. Schettler, publishing house Karl Siegismund, Leipzig, 1915)

5

Captain Mosel is riding:

"Yes, yes, my boy, you're probably not used to the war either. Come on, stay here, you can't buck. No, no, around the corner to the left. So, good gallop. But if you break your leg, it's not my fault; remember that. Golly, that's a decent caliber. Bang, there he sits. Bloody hell, it's making smoke. Easy, easy, you're all sweaty. Ow, on a whim. Well, you see, there's the railroad embankment. Oh, Mademoiselle Lilli's house is on fire, too, thank God. I'd like to know what kind of pleasure these bengels take in burning the roof over their own countrymen's heads—heaven forbid! Nah, we're not getting anywhere here. Now, wait a minute, come on, be quiet. Yes, you are a beautiful horse. So, be quiet; we'll stay here behind the gable until the monkeys have pointed their ridiculous guns elsewhere. There's another shot. He went between the telephone wires. Oh, and there were still onions in the garden. Yes, yes, c'est la guerre. Jesus Christ, where's Steuwer? After all, or were they still mopping him up? Lieutenant Steuwer! Don't be so nervous. Oh, you get used to it. Don't bang your head like that. So, come on, let's have a good gallop to the battery. Open field. Maybe it's better to crawl in the ditch of the road after all. Dear God, of course, it's gas. It looks disgusting. But now we really must hurry. If only the men had tied the protective gauze to their snouts. For God's sake, the cloud is already up to the Sixth Battery. Poor boys. I can't understand why the infantry doesn't have reserve troops here. With a few machine guns, you could run the show here. The gas poisonous? That would be a mess. Come on now. No, you mustn't buck like that. That's it, head straight ahead, see, that's a decent gallop. Christ, the houses in the firing position

are on fire. Yes, my boy, and if it costs you your life, it won't help. And the cloud is getting closer. If only we knew what that disgusting yellow stuff is. Respirators, will I put mine on myself? Too silly, after all. Well, where's the train attendant's cottage? Gosh, blown away. Trot, no one can get through here. Look, the scavenger gang is shooting with machine guns. No, wait a minute, it's dangerous here. Come on. Don't be afraid. We'll be there in a minute. Golly, those monkeys always have to shoot at the road. Should I let the horse go? Steuwer? He's not usually afraid, is he? Aha! Well, he was lucky. It certainly wasn't a meter in front of him. Steuwer, look, with a mustache tie. The chicks are always smarter than the hen. For heaven's sake, it's almost as if the Frenchman is aiming precisely. And how it stinks. The infantrymen call it Krakauer.[10] Kra-kauer, actually a very nice word. Kra-kauer. Poetry of war, silly, besides, we really have other things to worry about. Yes, dear Fritz, we'll split up here, if the gentlemen follow, they'll catch you. I'm sorry. Goodbye. Not very nice, really. I would really like to know what such a horse is thinking during the war. And now even a gas attack. Gas attack. God yes, gas attack. The Sixth Battery is no longer firing. Yes, that is unthinkable. Good thing mine is firing.

"Twelve hundred, if that's enough. But Fricke is no donkey. Twelve hundred, twelve hundred, that was a quarter of an hour ago, or half an hour ago, but if the Frenchman can run, then he could . . . Stupid dogs! If anyone sees me now, what can you do? Keep your nose in the dirt, better than, well, thank God, we'll sort it out in a minute. Where's Lieutenant Fricke, guard? What a shame, the poor guy. One of the old ones. Well, he's better off than us now, has no more worries. But pale in the face. We mustn't forget to take off his boots later. The sergeant's house is on fire too. Where is Lieutenant Fricke?"

6

Mosel leaps over the dead guard in a single bound.

A glance to the right, to the left. The house above the telephone shelter has been shot down to the foundation wall.

---

[10] Krakauer. German name for a smoked sausage from the town of Kraków / Poland.

But otherwise? Thank God, the battery is still shooting rapid fire. Sergeant Burghardt arrives. "Herr Hauptmann, we're firing at eight hundred; I've given the command myself. The telephone shelter with Lieutenant Fricke and Gunner Reisiger is buried. I have no men to dig it up."

He makes an embarrassed gesture. Mosel understands.

Off with the horse, let it go wherever it wants.

And upright along the guns through the firing position.

The crew of the left wing gun is complete, firing, not to be disturbed.

The next gun fires more slowly. The sergeant sits on the gunlayer's seat, an old soldier on the loader's seat. The two men are operating the gun completely alone. The other three are lying in a heap next to the gun carriage.

Mosel steps carefully over them.

The Fifth and Sixth Guns appear to be complete. No. Sergeant Lahne is lying next to the wheel of the Sixth Gun with his body torn open. But they are firing.

The First and Second Guns are in order.

"Sergeant, we can't keep firing at eight hundred without an observation post. Don't you have a connection?"

Burghardt does not answer. Figures appear half right in front of the battery, waving white cloth vigorously. "Herr Hauptmann!"

Mosel pulls the binoculars up to his eyes: they are Germans, approaching in long leaps. After a while, they are German artillerymen.

"Battery cease fire!" Yes, for God's sake, are we shooting at our own soldiers?

"Herr Hauptmann, these are gunners from the Sixth Battery."

A sergeant rushes out of the group of waving people. "Herr Hauptmann, the enemy has broken through. We're the only ones he hasn't caught. The gas is to blame for everything."

The gas? Oh, damn it, the gas! Now Mosel sees that a white cloud has come within four hundred meters of the firing position.

"Gun leaders, get the guns on the road immediately!"

How is this to be done? The wheels are feet deep in the rubble, with the dead lying between them.

But the Sixth Battery has been captured and the gas cloud is moving, and it must succeed.

"All crews to a single gun! Pull with long ropes, clear the way with pickaxes!"

The gas cloud is getting closer!

It succeeds. Only the Third Gun won't budge. It stands as if encased in concrete between the remains of the collapsed house.

They try again and again. Burghardt and Mosel have just joined the rope. Then, a pelting hail hits the shield. Machine gun!

Machine gun, machine gun? Then, the enemy has not only broken through at the front, but in a few minutes, he must burst the gas cloud and fall into the firing position.

Everything lies flat on the ground. It buzzes over them and smacks on the stone rubble. For a while longer, for a while longer, cursing this buzzing.

Then Lieutenant Steuwer turns the corner on the right wing, panting. He crawls up on all fours. "Just in time," Mosel shouts through his arched hands.

"I'm moving off toward Lens with the guns we've got out. You see to it that the cannon here is destroyed. Everything follows me, except for three men who, with the lieutenant, will wreck this place."

They scamper off like rabbits, one after the other, disappearing.

Steuwer hits the barrel of the gun with a pickaxe. The pickaxe bends. He turns blue with rage. Then, he pulls out the breech with the help of God. He tugs at the breech and smashes it into the dirt. The gunners shovel sand and stones into the barrel. So, well taken care of. No more shots will come out of *that* muzzle.

And off they go after the others. They jump over the rubble of the houses and stand in the street, now covered.

All the houses are burning. The shattered window glass looks like gold paper. Pig of a mess, goodbye bed and knapsack and blanket and dinner service.

At the last gable on the right, the Protzen's dispatch rider holds the horse by the halter. Shrapnel bursts above them. The horse puts its head between its legs. Before the white smoke clears, the second shot rings out. Here, as long as the houses are still standing, everything is reasonably well protected against shrapnel. The battery huddles together in a herd.

Third, fourth shot.

Now, it's getting too much for the captain. "We have to shoot! Gun leaders!"

And as they kneel around him: "The gunners are acting independently! We must get to fire. Line up on the road to Lens, covered by houses if possible. Put as much ammunition on the mounts as you can carry!"

So once again, in front of the burning position. Ammunition baskets move from hand to hand. Then: "Gunners to the guns! Push off!"

The guns move in.

Mosel and Steuwer in front. "Get the Protzen," the captain yells to the dispatch rider. He gallops off at a stretch.

The two officers follow behind him. They run at a crouch. As they reach the street toward Lens, detonations howl everywhere. They lay down and run until they finally reach a hallway. Cover. "How are the guns supposed to get here?"

Mosel looks up the street.

It works. A first gun is already turning into the street. A gunner carries the tail of the gun carriage on his shoulder; the others push at the seats and wheel scars. Bravo. Nobody thinks about the enemy fire. "Steuwer, the battery is magnificent!"

And now the other guns. Quick, quick, a few hundred paces, then we'll show that we're still here!

The leader of the first gun jumps up. Blood pours from his mouth. Too bad.

But the others give him a wide berth and come closer.

The two officers have left their cover and are walking beside them.

The shrapnel is much too far away and too high. They are harmless.

Infantrymen from Lens come toward them at a charging pace. Two long lines, pressed against the houses, rifles ready to fire.

They pay no attention to each other.

Finally: "Battery halt. Now, boys, you can fire!"

Two guns are lined up on the right side of the street, three on the left, closely staggered. Mosel is crouching at one gunlayer's seat; Steuwer is over there with the other group.

All eyes are watching.

The enemy's artillery fire has disappeared from the road. It seems to be concentrated around the church of Lens. From there, it barks incessantly.

The silence in the forefield is agonizing. Something must be happening. The gas must become visible. Or our infantry that had just advanced must come back. Or the enemy must finally appear.

Nothing happens. Waiting.

The battery waits, a minute is an hour. Waits ten minutes.

Then the gas is there. Yes, is it the gas? A thin fog creeps into the mouth of the road.

Mosel pulls up the binoculars. Shouts: "Put on your breathing protection!"

The gauze bandages are tied on.

Is that the gas? All eyes are glued to the cloud.

It is restless. It sways, swings, spills up in places, like a curtain in the wind. It tears apart.

"Herr Hauptmann!" Crowds of people appear.

Is our infantry coming back? Scattered? Then it will probably be—

Mosel still has the glass to his eyes. Now he shouts, boiling, "Boys, it's the enemy! Englishmen! Shoot low . . . but only shoot if you can hit them properly, save ammunition!"

Aim, aim, aim!

"Shoot!"

A whole group of people fly into the air—from the first shot.

All eyes are glued to this group. One Englishman is still on his knees. You can clearly see that he is getting up. Then, as he tries to jump back, he falls face down on the stones.

A new group is visible.

Now, a chase begins. Shoot! Shoot!

The guns become jealous of each other. "Please, you already have; now it's our turn." Shoot!

Nevertheless, the enemy does not stop. The Englishmen seem stunned and don't think about firing themselves.

A modern slaughterhouse: Cattle are driven into a lane, wide at the beginning, then narrower, then to their deaths, and none of them can go back because the crowd is behind them.

And always new dense groups, always new. Now, those at the front finally see their dead comrades in front of them and want to give way. But behind them is the crowd. The wall across the road gets higher.

The 1/96 becomes livelier from shot to shot.

No cover is needed here, no caution, no fear; everything is treated like on the parade ground: infantry advancing straight ahead!

It is inconceivable that the target, living people, can be shot down without any resistance.

At last, the enemy has realized where death is coming from. From the half-left, from the direction of the old firing position, sharp machine gun fire hits the battery.

Flank fire! Now, the gunners are lying on the ground. If only we knew where the M.G. is!

Mosel crawls forward. He tries to raise his head several times. But the bullets shave the terrain, making it impossible to raise the binoculars to his eyes. "Everybody take cover! Don't fire anymore!"

And then a second M.G. whines. There, from the right, a third.

Mosel can't even take the few steps backward to the First Gun. They are pinned down.

Waiting.

Someone at the Third Gun suddenly yells, "Mother!"

You can't even help him. He flails around wildly, and you would have to bandage him up.

Everything seems nailed down. Whoever raises his head has had enough forever.

Waiting.

Suddenly noise. From Lens. Mosel looks back. For heaven's sake, the Protzen are coming. If they get between the M.G.s here, it's all over.

The Protzen? Yes. The rattling is loud enough for everyone to hear. There's only one thing to do to try and save the day: pull back each gun one by one. If you die, you die.

Command. The shouting continues, "Back! Towards the Protzen."

And, if possible, wave the handkerchief carefully so that Hollert notices.

Slowly, like fat turtles, the guns move backward. It tugs at all the wheels, with outstretched arms, jerking heavily, by centimeters. And Steuwer meanders, a white rag between his teeth, nose almost in the gutter. And gives signals.

"Hold the Protzen!"

If only the M.G.'s sheaves didn't move with them.

No.

Battery back, by centimeters. Then, as the heavy guns started to roll, lively force meter by meter. And quickly more. And quickly further and further backward.

Protruding against the houses, two on the right, three on the left.

The M.G.s kick up dust, splinters of stone, always in the same place. And "Limber!!! To ... backward. Battery trot! Trot!"

The regimental commander comes around the bend in the main road, with his staff. Mosel reports, also for the rest of the Sixth Battery.

Brief consideration: "The 1/96 takes up position there in the garden, covered. Observation if possible at the height of the former position ... Await!"

The captain passes the order on to Steuwer. There is the garden, behind a wooden fence. It is knocked down, battery gallops in, guns take cover, protrude back onto the road to Annay. "Establish telephone connection!"

Nothing more to be seen of the enemy. He is completely silent. The gas cloud is gone.

"Too bad about the Sixth," says the commander. "The regiment has bled a lot in the last few days ... Do you think we can save the guns? See to it tonight, Herr Hauptmann Mosel."

"By your command, Herr Oberstleutnant."

"And find a lookout immediately. And have a cable laid to the staff quarters. Behind the church, in the former pioneer's casino."

He disappears.

Mosel rides after the battery.

7

General Staff 3rd Bureau
No. 8.565
14. IX. 1915

Secret.

To the commanding generals:

The spirit of the troops and their spirit of sacrifice form the most important bond of the attack. The French soldier fights all the more bravely, the better he understands the importance of the offensive actions in which he is involved, and the more confidence he has in the measures taken by the leaders. It is, therefore, necessary for officers of all ranks to inform their subordinates as of today of the favorable conditions under which the next attack by the French forces will take place. The following points must be known to all:

1. To attack in the French theater of war is a necessity for us in order to drive the Germans out of France. We will both liberate our compatriots, who have been subjugated for twelve months, and wrest from the enemy the valuable possession of our occupied territories. Moreover, a brilliant victory over the Germans will persuade the neutral nations to decide in our favor and force the enemy to slow down his action against the Russian army in order to meet our attacks.

2. Everything has been done so that this attack can be undertaken with considerable forces and enormous material resources. The uninterrupted increase in the value of the defenses in the first place, the ever-greater use of territorial troops at the front, the increase in the English forces landed in France have enabled the commander-in-chief to withdraw a large number of divisions from the front and keep them ready for the attack, the strength of which is equal to that of several armies. These forces, as well as those held in the front, have new and complete means of war. The number of machine guns has more than doubled. The field guns, which have been replaced by new guns as they wear out, have a considerable supply of ammunition. The number of motorcades has been increased, both for supplying and moving troops. Heavy artillery, the most important means of attack, has been the object of considerable effort. A considerable number of heavy-caliber batteries have been assembled and prepared for the next offensive operations. The daily rate of ammunition intended for each gun exceeds the largest consumption ever recorded.

3. The present time is particularly favorable for a general attack. On the one hand, the Kitchener armies have completed their landing in France, and on the other, the Germans have withdrawn forces from our front just recently to use them on the Russian front. The Germans have only very meager reserves behind the thin line of their trench position.

4. The attack is to be a general one. It will consist of several large and simultaneous attacks on very large fronts. The British troops will participate with significant forces. Belgian troops will also take part in the attack. As soon as the enemy has been shaken, the troops on the parts of the front that have remained inactive up to that point will attack to complete the disorder and disintegrate the enemy. It will not be a question for all the attacking troops of only taking the first enemy trenches, but of pushing through day and night without rest, over the second and third lines into the open territory. The entire cavalry will take part in these attacks to exploit the success with a wide gap in front of the infantry. The simultaneity of the attacks, their force, and extent will prevent the enemy from concentrating his infantry and artillery reserves on one point, as he was able to do north of Arras. These circumstances will ensure success.

The announcement of these communications to the troops will not fail to raise the spirit of the troops to the height of the sacrifices demanded of them. It is, therefore, imperative that the communication be made with wisdom and conviction.

Joffre

8

Lieutenant Fricke and Reisiger are sitting in the telephone cellar of the abandoned position 1/96.

They tried again and again to get out, to free themselves. They did not succeed. The cellar door to the courtyard is blocked, the frame is locked, and the weight of the shattered house fills the stairs. The hole toward the street is also buried several meters deep. A futile effort.

There is only one consolation, a weak one. Somewhere in the rubble, there must be a gap—from the street, an arm-thick bundle of light falls into their prison. Or no, no consolation. You go mad with rage by seeing daylight without getting out.

Anger gradually turns into despair. Trapped hopelessly. Actively trapped down here. *And how will it end?* It's a straightforward calculation: starve to death, or, if all goes well, wait for a second shot to put a quick end to it.

Fricke has been pacing back and forth for half an hour. He has his hands in his trouser pockets. Back and forth. With his foot against the cellar door, with his foot against the wall facing the street. Back and forth.

Finally, he sits down on the floor in a corner. Crosses his legs, stares in front of him.

Reisiger wants to alleviate the desolation of his imprisonment. He creates festive lighting. He puts candles all over the table, the entire supply. In the end, sixteen little flames are lit. The room is radiantly bright. Now, at least, you can no longer see the daylight. Then Reisiger trots. Trots, hands in front of his chest, trots up, trots down. If only there wasn't dead silence everywhere.

He steps up to the light shaft and puts his ear to the opening: lively machine gun fire.

What might be going on out there?

He flinches. *Who's saying something? Oh, the lieutenant. I completely forgot.*

"Reisiger, why are you doing this stupid lighting? It looks like a funeral."

Reisiger extinguishes the candles.

As only two are left burning . . . "Nah, Reisiger, it's even bleaker this way. You'd better light the plunder again."

This is done.

Fricke stands up. "How do you imagine this mess will continue?"

Reisiger stands at attention. "I don't know, Herr Leutnant."

"It's not clear to me that you can't hear anything from our battery?"

"By your command, Herr Leutnant. Doesn't Herr Leutnant think they're still firing? You just can't hear anything down here."

Fricke roars out with animal fury, "That's a mess! We're sitting here like monkeys in a cage, and who knows what's happening up there."

"Did Herr Leutnant see the gas?"

Gas? Then the battery could have been overrun long ago. *Perhaps,* Fricke thinks, *we are now sitting between two fires. All we need now is for the enemy to throw a hand grenade through the light shaft, and then good night.*

"Yes, Reisiger," he says matter-of-factly, "maybe it's best if we shoot ourselves. We'll never get out of this place. So why do we want to leave the pleasure to the Frenchmen?"

For Reisiger, Fricke is the epitome of a good officer. No one is more dashing, which makes the suggestion all the more disconcerting.

That means things are worse than I could have imagined.

Reisiger looks at Fricke suspiciously from the side. He stares into the candle and plays with his revolver pocket. "Well, Reisiger, what do you think? It really is . . ."

He doesn't speak any further. Footsteps are clearly audible in front of the light shaft. "Well, there's someone standing there, isn't there?" He looks up.

Bounces back, whispers, "Brown wraparounds, by golly!"

He pushes Reisiger into the corner and knocks out the candles. "They, they're Tommis. Now I don't understand the whole thing anymore." He leans against the wall and sighs.

Reisiger's knees tremble.

Thoughts: *What on earth is going on, and above all, what can we do? If we shout now, we'll probably get our muzzles stuffed with a hand grenade. If we don't shout, we'll get behind enemy lines and slowly die. But there is a third possibility: that the enemy is knocked out of the position again. That could be our salvation.*

So there's nothing left to do but wait and wait, leaning quietly against the wall.

From time to time, Fricke creeps back to the light shaft. Nothing more to see, nothing more to hear.

One can only note that the sun is shining.

Later, that it is setting with a red beam of light.

Later, that it is now dark.

Fricke and Reisiger are overcome with fatigue. But it is impossible to sleep. Stones and mortar keep falling from the wall. And that makes them wide awake again and again.

9

Divisional Order of the Guards Division:

On the eve of the greatest battle of all time, the Commander of the Guards Division wishes his troops good luck. He has nothing to add to the cheering words of the Commanding General from this morning. But let everyone keep two things in mind:

1. that the fate of future English generations depends on the outcome of this battle,

2. that great things are expected of the Guards Division.

As a Guardsman with over thirty years of service, he knows that he has nothing more to add.

Lord Cavan

10

Where is the enemy now, twelve hours after the beginning of his attack, today, September 25, 1915, four o'clock in the afternoon?

Great guesswork in the new firing position of 1/96. Everything happened so quickly that no one really understood what was going on. Gas? Yes, of course, there's nothing to talk about. Everyone saw the big gray clouds that were driven into the position by the wind, how the thick swathes ate up the Sixth Battery. But what else? For example, did the infantry suffocate in the gas and was simply overrun?

There was lively discussion at all the guns.

The only tangible thing had been the English, who had been gunned down from the road. Everything else was completely unclear. And above all, the vital question remained unanswered: Is there still German infantry ahead of us or not? Or, more specifically, could it happen to us that the enemy would suddenly appear here next to the firing position and plunge his bayonet into our stomachs, or would he be stopped in front? The infantry went ahead; everyone saw that. But if you think about it now, it could have been a battalion at most. And is that enough protection?

It's actually incomprehensible why the enemy didn't fire. Nothing, not a single shot. Not even beyond the firing position into the town.

Wishes were expressed, as well as curiosity and concern: One would like to go back to the old firing position to see what was actually going on in front.

Stupid situation. The captain had disappeared into one of the houses with Steuwer. So, there was no news from them. All the more, the attention was focused on Sergeant Burghardt. He was sitting on the gunlayer's seat of the Fourth Gun, leaning back comfortably, with the non-commissioned officers standing in a circle around him. He talked incessantly. Every now and then, you could catch a word.

He acted very important: "The enemy must have broken through, otherwise no Englishmen could have appeared next to the old firing position. But then the idiots seem to have lost their courage, and our battery has shown them how to fight back. I reckon they're now sitting in our cellars and eating our canned meat. It's a good thing the houses above their heads have burned down; otherwise, some bitch would surely be lying in my bed."

The enemy in the cellars of the former firing position?

The crews craned their necks. "If that's true," Aufricht said to Georgi, "then we won't even be able to bury Reisiger and Fricke."

It occurred to them for the first time that all sorts of people were missing. If you're dead, you're dead; there's little to say about that, and if you can't bury the dead right away, they're not impatient. But perhaps—Georgi put on his helmet—perhaps Fricke and Reisiger didn't fall at all. It's not the first time that people have had to live for days in a bullet-ridden dugout.

Georgi took Aufricht by the hand. "You, we'll get them."

The sergeant thought they were crazy. He flatly refused. It was daylight. "At least wait until evening. Or ask the captain, for all I care."

They went to Mosel. He looked at Steuwer. "Is there any point, Lieutenant?"

Steuwer threw away his cigarette: "If you don't mind, Herr Hauptmann, I'll go with them."

"Fine, but then try to find out where the enemy and our own infantry are. Or wait, I'll ask the regiment what they think."

Telephone contact with the regimental adjutant: "Tell me, Linnemann, what's going on? My battery has been here for three hours, and nothing is happening... How? Attack? Like this: I wanted to send Lieutenant Steuwer forward. Fricke is missing... All right! I'll report immediately what Steuwer has found out. How are the other batteries? Yes? Yes, the Sixth is gone. I had to see that myself. So? All the others before Souchez. Well, I don't think the mess there is any bigger than ours. Thank you very much. God bless you."

To Steuwer: "Well, you can go. But be back in an hour. Linnemann just told me that we're planning a counterattack in the section tonight at 7:30. By the way, the regiment considers the situation harmless... Well, off you go."

11

Where was the enemy?

The enemy had overrun the infantry lines behind the gas clouds this morning. Most companies fell in close combat. The others ran away when they saw the gas and were then destroyed by shrapnel fire.

The enemy pushed on. His columns stood on the ruins of the burning firing position a quarter of an hour after 1/96 had left it. There, for incomprehensible reasons, they stopped. From there, for incomprehensible reasons, the enemy withdrew to the front trenches shortly afterward, called back by command of his own officers. The direct fire from the 1/96 guns positioned on the road to Lens must have been the decisive factor. *Or was it?* Nothing is more senseless than chance!

In any case, the advancing replacement troops of the German infantry reserve reached the former 1/96 firing position without many losses and took up position there for the time being.

Steuwer, Aufricht, and Georgi appeared among this infantry.

A machine gun group was positioned above the bullet-ridden cellar of the former telephone shelter.

The position was in full view of the enemy. It was enough to raise his head to provoke one of the Englishman's machine guns to bark.

Steuwer was not deterred by this. He used his hands to tear open the light shaft to the telephone cellar. The gunners helped until their fingers bled. Shouting down the shaft . . . No answer from below. *Nevertheless, beams away, sandbags away! Into the dugout! Light a match!*

Fricke and Reisiger were lying there. Dead? When they grabbed Fricke by the shoulders and tried to pull him out onto the road, he began to rant terribly. They had woken him from his sleep. Surprise, shaking hands. "Thank God I'm getting out of this damned mousetrap! I hope the battery keeps firing bravely."

Steuwer explained. Fricke stuck his neck out of the cellar opening and saw the infantrymen. "Well then, I can give you the sacred assurance that the enemy took a walk here this morning. I have certainly seen English boots and English wraparound gaiters . . . Lord God yes, Reisiger is still here too."

Reisiger was lying under the table. They lifted him up.

He reacted like Fricke. He raged even more violently. He hit Georgi in the face with his fist and snarled. But then he quickly came to his senses and shook his comrades' hands: "Well, you have to be lucky!"

"If you don't give me something to eat now, I'll eat a child," growled Fricke.

"I've never been so hungry," whispered Reisiger.

Georgi reached into his pocket. "Herr Leutnant, if I can offer . . ." He had something black in his hand. It was Edam cheese, which he had been carrying with him for a few days.

Excellent! The dirt was scraped off with a knife, Fricke broke the piece and shared it with Reisiger. They ate like animals.

Then they left.

They walked single file along the houses to the new firing position. M.G. shots rang out! But they got through. The joy of the comrades was genuine. Nobody had believed that they were still alive. Only Mosel remained as a matter of fact. When Fricke came forward, he didn't even put down the newspaper. He said, "Fricke, you've let me down. Did you have to do that? All right then, let's get to work! We were supposed to counterattack tonight. It's been postponed until tomorrow. Time to be announced. In any case, the crews are sleeping at the guns, ready on alert. Lieutenant Steuwer will stay the night with me. You'll take an observation post on the chimney to the right of the road at the former firing position. Take Reisiger with you and another war volunteer. Have a telephone line laid to the battery immediately. Thank you very much."

A strange reception. Fricke was furious. He slammed the door rather violently, then looked at Reisiger kindly. "Yes," he said, "that doesn't help. But we've gotten used to each other quite well. Then we'll be able to spend the night together. Tell the sergeant that Aufricht is coming with you to the observation post. Take a telephone and something to eat. Or . . . there are to be rations from the Protzen . . . go to your gun. I'll send for you then."

Rations came. In other words, the Protzen and some of the wagons in the column brought ammunition and bread, jam, and cheese. Hollert was in the lead.

After reporting to Mosel, he had the battery assemble and read out the battery order under a flashlight:

"1. Sergeant Schulz and the gunners Schlichting and Kunz are going on leave on October 1.

2. It has happened on several occasions that soldiers have independently requisitioned food and clothing from the civilian inhabitants of Lens. This is strictly forbidden.

3. Gunner Reisiger is appointed supernumerary private.

4. The division's delousing facility is closed until October 1 for structural reasons.

Dismissed."

Lieutenant Fricke calls Reisiger and Aufricht and goes to the front with them.

According to the captain's orders, the observation point is the chimney at a defensive ditch near the old firing position. Reisiger and Aufricht tie a cable to the telephone of the firing position and unroll it to this point.

The chimney stands freely in the terrain. It has a door at its base. Fricke and the two gunners open it. This causes difficulties because infantry bullets are whistling from everywhere. Finally, they succeed: They are in the chimney. It's a black hole. When you light a match, you can see the black ring wall like a cage. Observe from here? You can only do that if you sit at the top of the chimney. "No," says Fricke, "I'm not crazy; if we try to crawl up there now, we'll break our necks. Put me through to the firing position."

Thank God, the connection works. Fricke tries to explain to Mosel how difficult it is to climb up the inside of a forty-meter-high chimney in the darkness.

Nothing can be done; Mosel is stubborn. He is obviously still annoyed that Fricke and Reisiger were not with him during the day. The gunners hear his sharp voice: "Official orders! Period!" So, it doesn't help. They light another match. They realize that the inner wall is bricked in with eyelets. Well then, you have to crawl up slowly.

For the time being, the telephone remains below with Aufricht. Fricke climbs, Reisiger follows. A hard piece of work. Less strenuous than unpleasant because it is so dark. Finally, they reach the top. They hang over the edge with their arms propped up, forty meters above the ground, forty meters above the field that holds friend and foe. An overwhelming sight. They forget that a misstep or the failure of their arms is enough to send them plummeting forty meters down the shaft. They forget that it is war. You just look.

You have a wide view. You see an enormous firework display. The horizon seems very close. There are flashes everywhere. Flames flare up everywhere. White balls of light arc finely against the black sky, ignite to form a large sun, and slowly float back to earth. Red rays shoot upward, becoming glowing balls. Toward Souchez, the whole earth burns.

Fricke is the first to speak: "Well, what do you say now, Reisiger? Is that beautiful?"

"Yes, Herr Leutnant . . ." Reisiger wants to continue . . . *It just shouldn't be war. But that seems inappropriate, or perhaps wrong, and he swallows the sentence.*

Fricke again: "Yes, Reisiger, then we'll establish ourselves here now. But how? The best thing is to crawl down first. Because a proper seat or something has to be created up here. After all, you can't spend twenty-four hours with your arms dangling over the balcony."

When they are back downstairs, he orders a carpenter and a bricklayer to be sent from the firing position with planks; the scissor scope is also to be brought up.

Toward morning, the observation post is ready. About one and a half meters below the edge of the chimney, a floor is set in, secure enough to carry three men. The scissor scope is positioned exactly above the edge.

As the sun rises, Fricke can see that the site provides an excellent view of the terrain below. He reports that the observation post is ready. After a short time, Mosel appears with two telephonists: "You are going back into the firing position. Our counterattack will take place at 7:10 a.m., without artillery preparation."

12

Fricke is reluctant to go back with them. Because now they see nothing of what is about to happen between the trenches.

But instead, they see something else, something very strange. The road they pass—the same one where they shot up the English yesterday—is teeming with civilians.

Where did they come from? Were they holed up in their cellars yesterday?

They stand, legs apart, hands behind their backs, scrubbed against the wall. They turn their eyes curiously toward the Germans. They pinch their lips mockingly, defiantly. Children laugh.

The facades of the houses have changed overnight. Flowers are tied to the hooks of the shutters, to door handles, to every nail. Asters and late phlox glow. Where there is a bullet hole in the wall, it has been planted and covered with greenery. Next to many a front door is a chair, carefully decorated with a white or colorful blanket. And there are flowers on it too, a pot of geraniums, a vase of roses.

What's going on now? What does it mean? "Strange magic," says Fricke. "Have you ever experienced anything like this?"

He approaches a group of women. He asks in good French what this is all about and points a finger at the flowers.

Many eyes look at him, bold and piercing and clearer than any civilian woman or man has looked at a soldier in months.

But their lips remain clenched. There is no answer.

The situation is embarrassing and shameful. Fricke says, "Pack" and walks on. A few houses behind, the next attempt. "Why the flowers? If you don't tell me, I'll hit you between the ears."

Smiles. But not a word.

Fricke pulls a bunch of white chrysanthemums from a window shutter and hits a girl across the face. Then he knocks a chair with his boot. The shards of a flower vase clatter to the floor.

Another smile. Not a word.

The soldier's anger is boiling. "Nothing to do with these bastards. Come on."

Fricke laughs, a stifled laugh. "Do you know what they're up to? The bastards think they can welcome their fellow countrymen here today with flowers. Come on, so we'll be at the battery when it starts." He spits at the windows.

In the position, the guns are ready to fire.

The battery is waiting.

Seven o'clock. 7:05, 7:09. In a minute, the infantry will try to conquer their old position.

Nobody needs to look at their watches anymore: Rifle fire starts. M.G.s yelp. Then you hear the fierce barking of hand grenades. The air trembles, bursts into thousands of screaming shreds.

If only there was something we could do!

If only we could see something!

So, the gunners lurk, paralyzed to speak, and listen ahead.

When will the captain give the command?

Fricke and Steuwer are standing on the right wing next to the telephone. They are not talking either.

The hail up ahead fluctuates in strength. One moment, it's shrill and loud, condensed to a single point; the next it's more ragged, scattered, stuttering.

Will it succeed?

Everyone has been in the field long enough to know that you can take a trench that is only three hundred to four hundred meters in front of you within ten minutes or never.

Ten minutes are long gone.

The fire does not let up. The restlessness in the battery increases. Everything is feverish: If only one shot could be fired! It doesn't matter where, just that the stupid tension is released.

Aha, Lieutenant Fricke picks up the telephone.

The eyes of the whole battery are on him.

A long conversation. Finally, he hangs up. He turns to Steuwer. Every word is understood up to the last gun. "The attack seems to have failed. The enemy trench has only been reached in a few places. We mustn't fire for the time being so as not to hit our own infantry."

Always the same rattling. Attack failed? Too bad, there's nothing we can do. Gradually, the noise becomes boring.

Finally, you hear a new sound. *What is it?*

On the street, visible between two houses, new infantry is advancing at a run, rifles half raised. Reserves.

The sight is paralyzing. So, the battalions that were deployed this morning are finished?

Reisiger looks at the panting people. Many men with beards, among them very young officers with pistols in their hands. Poor dogs. He can't help thinking of the civilians on the street with the flowers in the windows and in front of the doors. Hopefully, the officers will find enough time to sweep away the goddamn garbage.

There's a roar in the air. An older officer throws up his arms and slumps backward. Six infantrymen lie down, twitch, and stretch. The pursuing soldiers run past them.

The enemy is firing shrapnel! It begins to scatter. White clouds billow above the houses. At least the infantry is through now. Now you have time to pay attention to other things again.

The gunfire? Reisiger nudges his neighbor:

"Do you hear? Silence of the grave."

"If the captain says that the counterattack has failed . . . Who's going to shoot then! Man, we artillerymen have it better than the foot soldiers."

Yes, Reisiger thinks, *we have it better. A few pieces of shrapnel here . . . and in front of us, the infantry is up in arms against M.G.s.*

He looks at Fricke. He is just putting the phone down again. "Listen up, everyone! The matter has come to a standstill for the time being. The battery is dismissed for now!"

It's come to a standstill for the time being? So, it's gone wrong.

"A mess."

The neighbor doesn't understand the context. "I'd rather have lunch soon too."

Hours pass. The field kitchen rattles up. There is bean soup. After the meal, the gunners take the playing cards out of their skirt pockets. There is mumbling everywhere. Fricke and Steuwer stroll around playing lapwing. Reisiger, who knows nothing about playing cards, loses last month's pay in no time. "You're getting a bonus now, Herr Gefreiter,"[11] one of the winners laughs at him. Everyone grumbles. "You, that'll cost another round. Wait till we're in some peace and quiet."

A sergeant has two buttons in his pannier. "You'll have to sew them onto your jacket collar now!"

Five minutes later, they are on his collar: Gefreiter Adolf Reisiger.

What you can't become!

Reisiger thinks that more than half of the war volunteers who went into the field with him have already become non-commissioned officers, and half a dozen have already become lieutenants.

He no longer enjoys playing cards.

Toward evening, the captain decides that no officer is needed for observation tonight. One sergeant and two men are enough.

When it's about eight o'clock, there's bread, cheese, tinned liver sausage, and two large pots of hot coffee at the field post. Later, we curl up in our blankets. A cold night, not good for sleeping. Nobody wants to talk either.

A light is on in one of the houses on the left. The captain is sitting there with the lieutenants. "It's a long time since I've seen such a mess

---

[11] Gefreiter. In World War I, Gefreiter was the only rank that conscripted soldiers could be promoted to. Equivalent to a private first class or a lance corporal. The rank insignia was a button, the so-called Gefreitenknopf, on each side of the collar.

as this morning; it wasn't possible to get the pack out of the trench. The question is how to proceed. Fricke, from four o'clock tomorrow morning I ask you to take over the observation again with Reisiger and Aufricht."

13

The night is not quiet. Noises cut long furrows through the darkness everywhere. Gunshots. Shouts. Whispers. Rumbling. And footsteps. Marching footsteps, nonstop, monotonous, with shuffling heels. Reisiger stretches; all the sounds pass through him, over him. Footsteps, columns. It's moving to the front; it's gathering here. Tomorrow, it will rise.

He thinks of the civilians. In view of the threats of this night, they are no longer imaginable. When the sun rises, it will be soldier after soldier awaiting the enemy. Good thing the flowers will disappear then!

Flowers. That's what his thoughts are clinging to now. That the women's gesture is beautiful, and their pride too, and their senseless certainty that they will win.

Then he remembers the girl in Annay. Marie. *Is she asleep? Is she crouching at the entrance to the village, next to the old woman, waiting too?* When Fricke calls him at three o'clock, he is still awake.

3.30 a.m. They are already standing on the platform of the chimney with Aufricht.

The sky is colorless. To the east (Reisiger thinks, *Germany*) it is a smooth yellow. No clouds.

"If it rains early this morning, it's half the battle," says Fricke, sniffing. "Well, we'll manage to sort things out."

The phone buzzes. "This is observation 1/96."

Order from the regiment: "Red-green flares in the section mean gas attack." Reisiger repeats half aloud.

Oh, they'd forgotten that. Right, there was a gas attack, or rather, there had been a gas attack. Today again?

Fricke has the battery made ready to fire and directs the firing targets. Reisiger and Aufricht look over the edge of the chimney with him. There, in the middle of the field, flames go up, followed by a crash. "Those are our guns."

The white limestone edges of the position taken by the enemy are faintly visible.

New order to fire! Fricke places the shots so that they flank two approach trenches.

The enemy artillery does not respond. "Halt! Cease fire!"

Mosel calls . . . *Is there anything new? No.* This is repeated several times now.

It has now become so bright that you can see deep into the country. Up ahead lies Loos. A small, white farmstead lies huddled in a thick hedge. There is a black ridge: the coal heap. *A few days ago*, Reisiger thinks, *I was still standing up there and now the Tommy will sleep in it*. In the distance, the houses and towers of Grenay. Floating like ships with red sails on the horizon. There is a trough in front of it, a hollow. You can't see the bottom even from the chimney.

Fricke sniffs at the sky again. The sun is rising, a thick red ball. "It's going to rain, too bad."

Cold wind from the direction of the enemy.

Oh, wind from the enemy? Fricke presses his eyes against the scissor scope. "Wind from the enemy," he says aloud.

He is about to turn away when Reisiger cries out. "Herr Leutnant . . . !" That's all he can say. Green-red balls of light dance before his eyes. The green is only faintly visible against the morning sky, making the red all the more threatening.

Green-red everywhere.

From Loos to the slag heap like a veil, green-red splashes up, dances, staggers. At the same time, an insane barrage of gunfire.

"Gas!" They all say it at the same time.

How different it looks from up here compared to the day before yesterday from the firing position. You can see with the naked eye that the gas cloud is rising from the former first enemy line. It is drifting forward very quickly. The English are lined up in the old German infantry position, guns at the ready, bayonets fixed. Steel helmets. A yellowish muzzle in front of their faces. They have gas masks! All you can see are eyeglasses the size of your palm and a trunk.

An eerie picture. Nobody moves; they are obviously letting the gas cloud drift over them.

Telephone. Without thinking: "Battery to the guns! To the specified target. Enemy is making a gas attack!" The first shots from 1/96 hit the waiting Englishmen. Fricke has the guns swivel a few meters after each shot. He bursts into howls of joy when they score a direct hit. He doesn't care about the gas cloud. He keeps his eyes fixed on the scissor scope. "Golly, that was a good one!"

"We sent at least half a dozen of them to hell."

"Ha, you can really see the legs flying!"

"Thank God, other batteries are firing now."

Aufricht is sitting at the telephone, and Reisiger looks over the edge of the chimney.

The enemy is under heavy fire. The gas cloud makes no impression. It is thin; it is barely a man's height before it becomes as transparent as a flimsy fabric. The sun's rays shine through it in broad golden stripes. Now the cloud plunges into the hollow. It is swallowed up. So now the curtain is drawn back, and the view is clear.

14

Curtain up!

Curtain up, Fricke away from the scissor scope, his hand grabs Reisiger by the collar: "Cavalry!!!"

Cavalry. The enemy attacks with cavalry. Cavalry rises from the hollow. Shiny gray bubbles come: steel helmets, a chain, from Loos to the coal dump. There come bobbing horse heads, bobbing a chain brown and black and white, from Loos to the coal dump. There comes brown and black and white, a load densely massed, pressing on without gaps from Loos to the slag heap.

A roll of cavalry grows out of the hollow, pushed up and jumping up in a closed front! High, and now visible. And stands, incomprehensibly, slowly pausing, between the hail and rain and thunderstorms of the German infantry, the German batteries.

Fricke looks through the glass: "A second line is coming!"

The same to the naked eye. From the hollow, behind the slowly pausing mass of pushing horses, again horses and riders. From Loos to the slag heap.

"Herr Leutnant!" Aufricht has put down the telephone and is staring next to Reisiger.

Fricke: "Let them, let them ... they must come here."

He says this hoarsely, eerily, roaring. "They must come here." Reisiger's knees tremble: "There!"

There is still no artillery shot into the waiting, rising phalanx of horsemen.

And Reisiger, his arm suddenly under the lieutenant's arm: "They're riding up!"

They ride on. The first massif rises briefly. Lowers. Raises. Down. Up, down, up, down, up, trot. Behind it, the second, up, down, up, trot, simultaneously two limbs, tightly pressed, trot. And trot. And, lunging, a leap of the whole front, horse's legs reaching out, hooves stretched in the air and up and stretched down and canter and bellies on the ground, neck forward, charge! Two ranks up to the trenches. And the riders, the lances still half high and now lowered, charge.

An animal falls, one breaks its knee, one throws its head over backward, a rider rolls off, and one drags in the stirrup.

The charge approaches.

A gap opens up, four horses wide, in the first ranks. Six, eight, nine tip sideways and twitch in a flapping heap. Charge. Unrestrained, uncorrectable, unstoppable living force, charge. Closer, closer, closer. Over the position of the English infantry, forward, closer, closer. The three on the chimney gasp. Eyes back and forth, first link, second link, back and forth, riders, horses, first link, second link, nearer, nearer, charge.

And—

Fire between the whole thing!

How many telephones are buzzing in this second?

How many voices shout: "Cavalry between the trenches!"

How many commands, hard, metal in tone, make them stop, break off the distance, drive a barrage, a wall of piercing flames, just before the German positions?

"Cavalry between the trenches!" Every leap means getting closer by three meters. Closer, charge—

And that's where the halt begins.

Everything forgotten by the Germans, out of hole and trench, high from the cover, standing freehand, gaze burning in the horse's eyes, in the rider's eyes, the blood red from flickering pupils directed at them.

Sight... fire! Light machine guns, heavy ones, fire! Field artillery, fire!

And all greedy bites, biting into the thundering, dull, groaning, thick mass of the cavalry charge.

The three on the chimney, commands thrust forth, passed on hoarsely, and now hanging over the edge.

This drama of the most monstrous proportions! The first row and the second row, no longer divided now. They collide into each other, already overrun, already a limb, already too close to the intended movement. And it saws and stamps and squeezes and digs and eats.

Machine guns between the beating legs of the horses so that the jagged stumps shuffle over the ground, shrapnel in front of the chest, grenades under the belly, bundles of sulfur-yellow flames, columns of brown smoke, fountains of blood and intestines as thick as an arm, limbs, and hulls of men and animals hurled up. All this, as far as the massif stretches, from Loos to the slag heap.

The whole thing now crumbles into blocks, gaps between them. The blocks, still ponderous in their forward pressure, break, pile up unevenly, and loosen, so that everywhere something jumps up, up, sinks, flails about, lies. All this: crushed horses, crushed riders from Loos to the dump.

And still no end. A group is still trying to pull the horses' necks around, away, back! And then something dashes off. And something crawls with fluttering movements.

Fricke, bellowing: "There, Reisiger, they want to break out, there . . ." And more bellowing: "Battery rapid fire, move up a hundred meters!" How many commands are chased through the wires: Rapid fire! Don't let anyone get away!

Rapid fire. How will even one rider escape?

The madness is awake, the last fear, the most horrible horror. Not a single horse turns. Still, the dead only press forward.

All batteries and every gun remain in front of them. A hundred collapse, again and again, and a hundred try to climb up, again and again. All the batteries, all the guns against it.

Even the dead are mangled again and again, again and again.

Hands rise from the tough, blood-covered rampart; faces, unrecognizable, rise; gestures flutter.

Standing freehand, the German infantry fire their shots ... until everything suffocates motionless in a mash of blood.

And the English infantry behind them, in the trench, separated from the Germans only by this steaming wall? Did they have to watch this, the destruction of their men, the death of their brothers down to the last man? Did they flee?

The three on the chimney, greedy: "Where is the English infantry?"

The telephone buzzes. Aufricht picks up the message. "Our infantry will storm the lost position in four minutes. All batteries: annihilation fire."

Fricke, calm: "Advance fire, old target!"

How many wires give the same command at the same moment?

The entire artillery of the section fires annihilation fire on the English infantry.

Fricke: "How many more minutes?"

Aufricht: "The assault will begin in one minute."

The English trench is plowed with destruction. And now the German infantry leaps forward, safe, wading through the blood swamp, up to the belt in slippery corpses.

Is the enemy firing?

There are some who tear the rifle to the cheek, and barely a single man across ten meters—they are hacked to pieces by grenades. No M.G., artillery very timid, only shrapnel, and the shots far too high to grab the attacker.

And the Germans, as their artillery drops the curtain of fire at the precise second, safely into the trench, into the old position.

They cut down with the bayonet who raises his arm. With hand grenades anyone who tries to climb the steps of the shelters by surprise. "The operation," say the telephone lines on the Loos to the coal heap section, "has been carried out as ordered." The three on the chimney are relieved around midday. The streets of Lens and the firing position are quiet.

The 1/96 has fired for four hours today without receiving a shot.

15

But to push through day and night without rest, over the second and third lines into the open terrain . . . These circumstances ensure success . . .

Chapter VIII

1

The regimental order of the next day, September 28, announced that 1/96 would be temporarily withdrawn from the regiment and would be at the sole disposal of the Army High Command as a so-called "flying battery."

At the same time, the battery was deprived of two operating guns for use in the formation of new reserve regiments.

The order came as a surprise. Something like this had never happened before, even to the "old warriors." The gossip began, and the rumor mill was churning.

Finally, everything came to a head shortly after the order was announced: A fine mess! Because "for special disposal" certainly doesn't mean that 1/96 is always pulled out where it's dicey; on the contrary! The clever ones explained it in detail: Wherever there is a mess in the next few weeks, we should get the car out of the shit.

Right, that's exactly what happened.

The captain was proud of this appointment.

After the mail was distributed, he gave an unusually long speech. The regimental commander had already called 1/96 "the Iron Battery" after the days in May '15, after all. This was new proof of the esteem in which 1/96 was held. Everyone should be proud and give their all to always maintain this glorious title. "You know that this is still one of the liveliest points on the Western Front. Even more reason for me to ask you to behave properly. You owe it to yourselves and to the comrades we have left in front . . . . The battery will move into its new position in front of Souchez this evening . . . . Stand still, dismissed."

Souchez? Reisiger thought of his dream. Souchez, Carency, Givenchy—it had been in the army report often enough in the last few weeks. *Hmm.*

The afternoon passed quietly. The team was charged with tension, irritability, and depression.

It was only when the Protzen arrived that the rattling of the guns and the snorting of the horses brought balance back into the mood.

The battery departed. It was not a long march; the new position was reached after an hour. A churchyard.

Within reach, a seething fire was boiling in a cauldron. Souchez was somewhere there.

Work began on extending the gun emplacements. As best they could in the dark, they dug holes to the right and left of the guns for the ammunition that was brought in.

Gravestones torn out of the ground, covered with turf, formed good ramparts in front of the guns.

There were no dwellings.

The men slept under the open sky. Mosel and Steuwer stayed with the battery for the night. Fricke was ordered to immediately seek a surveillance post in the infantry position and make telephone contact. Telephonist: Reisiger.

2

The 1/96 stayed here for ten days.

Reisiger kept a diary. A small oilcloth notebook with checkered gray paper. It was forbidden to make notes. But Reisiger helped himself by sending the written sheets home every eight days. This meant that they could not fall into the hands of the enemy even if they were captured.

His last notes:

I can't get out of the trench at all. We telephone operators are supposed to be relieved every twenty-four hours at the latest, but Fricke doesn't abide by any regulations. Sometimes I think he's not quite right. He does things that no longer have anything to do with so-called bravery, but which I simply call sinful recklessness.

This morning, he suddenly claimed that he would never be able to get a proper picture of the terrain here in the trench, that he would have to climb out (in broad daylight). Only then would he be able to see what was actually going on.

You can't say anything against that if you're only a Gefreiter. I had a stinking rage. I think the infantry thought we were both insane. He really did climb over the trench wall at about nine o'clock in the morning. I tried to hold him back, or at least to make him realize that I had to stay on the phone, but it didn't help. I had to follow him. I still didn't understand why that was necessary. Thank God, neither did the enemy. It must have been a strange sight for him when suddenly two life-size men were walking in front of the first trench. And not just for a few seconds, but even then, when several rifle shots were already flying around our ears and a company commander of our infantry was pleading with Fricke to take cover again, because otherwise the enemy artillery would inevitably be alerted.

Nothing to be done. Fricke stood with his legs apart, staring through the binoculars. I stayed behind him because I thought: If the enemy shoots, let him kill fat Fricke first, after all, it wasn't my idea.

And, of course, the enemy really did shoot. An M.G. It whined so unpleasantly. It was as if it had been set up right next to my ears. Then, finally, the lieutenant made himself comfortable to trot up slightly. He jumped into the trench. As I tried to jump after him, shrapnel hit the ground a few steps in front of me. I fell right between two dead bodies that were still lying in an old sapper from the previous night.

I angrily returned to the dugout with Fricke. I think he noticed. He kept making jokes, but I couldn't laugh.

In the afternoon, he practiced target fire with our battery in a completely unmotivated manner. Thank God, without incident, the enemy didn't respond. Of course, the infantry whined that we were always drawing the enemy artillery toward them with our cursed firing. They were right.

I heard a typical remark: "It's always like that, some artilleryman comes along and does pompous target practice and then goes home peacefully. And ten minutes later, we have to pay the price."

The mines here in the trench are very eerie. They look like bats. Sometimes, like giant bats. Today, for example, the Frenchman sent over some things that were a good meter long. You must learn to avoid them. You can see them coming. An infantryman told me: "You always have to keep one eye in the air and the other in the next shelter." I benefited a lot from that. Fricke was in the battery. I was alone and strolled through the trench. Nice weather. The infantry was playing peace. There was only one sentry every ten meters. The others were sitting on the trench wall with their shirts off and cracking lice. Suddenly, the enemy started with their mines. Heavy chunks, the first shot was right on target. The trench was almost leveled, thank God, without casualties. I was about to duck when I heard another short burst and saw the beast coming straight at me. I rolled into the next dugout. I hadn't paid attention to the fact that it had a lot of steps and plummeted down like a sack. At the bottom I was shouted at terribly. I had fallen against a lieutenant's table and tipped over his coffee pot. When I came out again, several infantrymen were standing in a semicircle at a breastwork. In front of them was a comrade with a finger-thick stream of blood coming from his neck. We bandaged him up, but he was already dying.

I think I experienced the worst night here in the field yesterday. Around nine o'clock in the evening, when I was sitting in our telephone hole with Rabs, an infantry officer came and demanded contact with Mosel. The enemy was very restless. He insisted that an artillery observer with a telephone should move to an advanced position. The infantry wanted to send patrols up to the enemy trench.

Mosel's order: "Rabs remains in the current dugout. Reisiger is at the disposal of the infantry. Communication must be maintained between him and Rabs." I felt as if I had been ejected from the battery. Of course, the infantrymen are also our comrades, but there is always something like enmity. So now I belonged to the "enemy" who "had" me.

The first French trench was about eighty meters from our line. We had advanced three forward trenches here. So did the enemy. The endpoint of our and the enemy forward trenches are only three meters apart.

I took a roll of wire, a telephone, a flare gun, a blanket, and cigarettes, and left. An infantryman led me. At first, the wall was a good man high. We walked under our wide wire entanglement. Then the approach became more difficult. We had to crawl on our stomachs. Finally, we reached the saphead. It was obviously an old grenade crater. There was a small barricade of sandbags facing the enemy. Some cheval-de-frise stood in front of it. "Of course you have to keep your mouth shut here," the infantryman whispered barely audibly in my ear. "The Frenchman across hears every word, every cough and sneeze. Generally, we don't do anything to each other. But if there's some crazy bastard over there, he'll throw a hand grenade over the balcony, and that'll be the end of it." The infantryman disappeared. I sat there. My first task had to be to find out if the telephone was working. That was quite a feat. If the enemy can hear the coughing, I thought to myself, he can also hear the telephone number and my speaking. What to do? I curled up like a hedgehog, put my head between my legs, pulled my blanket over me, and called. Rabs answered. "Connection established.... My dear Rabs, who knows if we'll meet again." Rabs laughed. I was about to explain my pessimism when I heard a loud cough, so I threw the phone down and pulled the blanket off my head. Yes, it was a cough. My heart was pounding. It was the post, the French post, the enemy in front of me.

I lay back in the crater and stared ahead of me. It's madness that two people are facing each other at night at a distance of three meters and, if they want to be decent soldiers, they must have no other concern than to kill the other person in every possible way and as quickly as possible. I had often thought about this, always with the same desperate result. If you don't kill, you will be killed. Or, if you don't kill him, he can kill one of your comrades.

Up here, all alone, this thought particularly depressed me. I realized, shamefully, that I was not a good soldier. I even toyed with the idea: how easy it would be to stand up now and shout "Monsieur." And then we would each walk a meter and a half and shake hands.

With a crash, a hand grenade flew to the side of me against our wire entanglements. All thoughts were chased away. Gunfire from our trench. Oh yes, we had sent patrols forward. Maybe the enemy noticed them and shot them dead. There's a war on in the end.

The Frenchman coughed again. A flare went up. Did it come from him? The flare was hovering above my hole on a parachute. It was a disgusting light. I saw my shadow, a deep black with a dark red edge on the wall of the hole. The glowing ball hovered lower and lower. When it was as high as a man above the ground, I felt myself painfully tearing open both eyelids. There was a human being within my grasp. I could clearly see his eyes, which were fixed on me, then I saw a steel helmet. The flare went out. I was blinded, and as a result, the darkness was even more impenetrable. But I knew for sure that there was a person in front of me. What could I do? I can't call the battery or Rabs. But I can't fire my pistol either, because then the sentry on the other side would surely throw a hand grenade.

Should I run away? I can't do that. Should I pull my blanket over me? Maybe so that the lurking man won't recognize me when the next flare comes up and sneaks past me? But the idea of not being able to see for myself seems unbearable. Should I shoot a flare myself? No, I can't do that either, because I only have two red ones; they mean barrage fire. And since we have patrols in front, our barrage would cause mischief among our own people.

My thoughts were running back and forth through my head. Finally, with no solution in sight, they paralyzed me. I lay down on my back and thought, so with God, let him put the gun between my eyes.

But this idea woke me up again. I found myself rolling up my skirt sleeves to my elbows. At the next flare, I'd jump at his throat and strangle him.

I couldn't. The next flare whistled up. Yes, there he was. Maybe, I thought, he'll get scared when he notices me and just runs away. So, I will try to lift myself a little over the edge of the crater and look at him with a jerk.

The flare was already halfway to the ground again. That's when I did it. I jerked up and stared at him. So close that I thought our heads were about to collide.

He didn't move. The flare was out. I felt horribly sick. I crawled back into the hole, knowing he was dead.

I have now seen many dead people, including many comrades. But this was unbearably horrible. Again, and again, flares and again and again the face. And each time, it became more familiar to me, and each time, more details stuck. Finally, I had the crazy thought that the dead man next to me looked exactly like me.

Now I was so far at the end of my tether that I wanted to jump up and just shout and run back.

Thank God the phone rang at that moment. Rabs: he was connecting with the battery. Lieutenant Fricke was on the line. Drunk. He had an oily voice and was full of a greasy benevolence. "Well, Reisiger, how are things going? You must be proud to be up there at the top of the battery under the beautiful starry sky. What's the news? Nothing at all. Well, all right then. At dawn you go back to the telephone shelter. Good guy, huh?" He hung up.

Good guy. Certainly not a good guy. But I was a little more composed. Finally, I forced myself to look at the dead man's face calmly. Yes, he really did look like me. I had shaved yesterday, and I had noticed how hard a face gradually becomes in war. Large areas around the cheekbones, deep-set eyes and the strange wrinkles that lead from the nose to the pinched mouth. That, I think, is where the war really ends, where it becomes so clear that man, the individual man, kills the individual man. Because he could be me, I could be him. Is there still any sense and any "enmity"?

After all, I was calm now, the usual thoughts came: Maybe he has it better than me, in any case he will not have been afraid of me tonight and he will not be afraid of those who sit here after me for the next few nights either.

A new flare. I hear footsteps behind me. Or maybe not a step, maybe the wire jangles? Nevertheless, I jump to the side. Then I see a black shadow gliding to the ground. Is that another dead body?

The flare goes lower, goes even lower, falls directly onto the black shadow, is about to go out—the shadow moves, strikes the glowing remains of the flare with its hand. But that is now . . . without doubt a human being!

I forget that the sentry is three meters in front of me and that the dead man is lying next to me, and I jump up. I know that I'm sitting on a crate containing hand grenades. Jumping up, I pull off the lid and have a grenade in my hand, for the first time in my life. It's a small mace. It has a thread with a thick glass bead at the bottom of the handle. I only know what you do with it from hearsay.

My eyes burn against the black shadow. Friend or foe? What should I do? If I call him, it may be too late, then his bayonet or his knife will be in my throat. If I don't call him? The watchword tonight, I know, is "Henthen." Coincidentally after the name of my hometown, because there are fellow countrymen in our section. I grip the hand grenade tighter in my right hand. The pearl between the thumb and forefinger of my left. I straighten up, and say "Henthen" quietly and questioningly. The shadow rises to its knees, and I can clearly see that it is making a turn toward the enemy wire entanglement. I snatch the bead from the stem and throw it after him. Two, three. There's a loud crash, a blue fire; I hear something like a scream, and I take the phone and blanket under my arm, flare gun in hand, and make a long leap back to the trench. There's a crash behind me. A blue flash shoots out of the hole I was just sitting in. Infantry rifle fire on both sides. In front of me, everything splashes onto the ground. I jump on, twenty steps, ten steps, I'm in the trench. A company commander charges at me. I tell him confusedly what has happened. He shouts "barrage." I fire two red flares from my pistol at the sky. Our battery begins. I jump into the bunker with Rabs. Phone buzzes: Report to Captain Mosel. All hell is breaking loose around us. I can still hear Mosel's calm voice: "Well behaved."

The infantry had very few casualties. Rabs and I were relieved at about half five o'clock in the morning. I had to tell the captain everything again; then I was allowed to rest. They had built a decent shelter out of gravestones under the roof of a family grave in front of the firing position. I slept wonderfully.

Two days later: The captain orders the battery to assemble in the firing position. He smiles at me and takes an object in blue wrapping paper out of his coat pocket. He unwraps it: the Iron Cross. He pulls it through my buttonhole on a ribbon that is far too long. Then he says a few friendly words, encounter with enemy patrols, or some such nonsense. I am quite amused. Father will be proud. I'm promoted to Unteroffizier that evening. That's more likable. I think that some of the physical drudgery will stop now. Sergeant Burghardt takes a greasy yellow braid out of his pocket that looks like a cigar band. I sew it myself to the corners of my collar. "They look really new," says Burghardt. But then he slaps me on the shoulder and even says "Comrade" to me.

God, I'm so proud.

3

The fighting around Souchez was heavy. Twenty-four hours never passed without the enemy trying to break through with superior forces. It did not succeed. But the fighting tore open the veins of the defenders; the section drowned in blood.

Alongside the assault from trench to trench, hand on the enemy's throat, alongside the artillery barrage, battery against battery, came a new horror: the swarm of planes. Six French against one German. Until then, they had laughed and mocked at every plane. Now it was doom.

Like vultures, the gray birds were on the lookout for victims. Poor artillery. It became hard to stay hidden, and easy to be crushed.

The battery 1/96 had daily casualties from accidental hits. Now, one night, when it had to fire a barrage because the infantry had the bayonet of the enemy on its neck, chance turned into correct calculation. An airman led the invisible command. And two heavy French batteries covered 1/96.

Mosel was in position. He tried to respond.

There were already seven dead and four wounded between his guns. He did not get a shot fired.

The men took cover against the churchyard wall. There they crouched, pressed against the wall, their faces turned pale against the burning night.

The enemy plowed the graves. They opened up. The dead came to life, skeletons sprang up, and they had to contort and twist under the pressure of the shells. A terrible hour!

Mosel tried to get a connection to the Army High Command: Staying here is nonsense. The telephone lines were shot up.

Contact Fricke, who was in the trench at the front. The lines were destroyed.

So wait until the morning.

And then, order: The battery stays in place at all costs!

The same thing every day. Dead men every day.

Material damage: Two guns rendered useless by direct hits.

Army High Command: Battery stays in place. Two new guns to be received immediately.

Again and again, again and again Mosel's evening report: "Fatal losses."

Army High Command: Too bad about every man. But there's nothing we can do. Good evening.

Finally, one morning, at dawn, a dispatch rider: "Change of position. Back to the Loos area."

As the battery drove through Lens, through smoking, shattered streets, past the leaning, torn-open cathedral, Reisiger had to think of his dream again: Shock: Souchez! Shock: Loos. "I want to drill these names into your head so that you won't forget them for the rest of your life. That would be something!"

And the glowing arrows moved across the map again.

4

If only it wasn't always the same! Souchez or Loos—war has become a machine, an automatic machine. Infantry attack: barrage. Artillery combat: response.

And response means losses. (And even silence means losses.) All gunners have only one thought: *When will we finally get out of this hell?*

The Gunner Reisiger has become the Sergeant Reisiger. It was the same: telephone in the trench. Or telephone in the battery.

And the army report, the closer it got to winter and the more it got cold and rainy: "Heavy fog hindered operations" or "In general, the noticeable slowdown that can be observed in the activity of our offensive must be attributed to the incessant downpours, which soften everything up and make operations almost impossible."

We read this; we heard it every day when it was passed on from position to position by telephone in the evening. The laughter at this became harder and harder, more and more bitter, and finally, of course: "But yes, there really are no 'operations' of any significance. The fact that our battery alone sends three or four dead to the rear with the food wagon every evening is indeed irrelevant. The fact that the infantry loses two or three men in the section every hour, day and night, is also not news for the army report. We have enough human material for that.

Only one thought offered support: Christmas is coming.

And there are so many childlike minds and so much belief in Christmas that suddenly the rumor spreads and is whispered on, from the sappers in front of the infantry position through all the tangled trenches, through all the approach routes to the field artillery batteries, to the heavy guns behind them—perhaps even to the General Staff: Watch out, there is peace at Christmas!

5

In the last few months, fraternal peace efforts have made themselves felt in Germany on several occasions, which require close monitoring. The supporters and promoters of this movement are mostly people of little political influence. They generally remained confined to the circles of "pacifists," who were already pursuing a vague cosmopolitanism before the war.

Given the determined, patriotic attitude of the German people, it is hardly to be expected that the movement will penetrate broad sections of the population and achieve decisive significance. Its toleration at the present time, however, must rightly give rise to dissatisfaction and opposition in wide circles and may ultimately impair the firm, unwavering will to persevere. It is not understood that the discussion of practical, patriotic war aims is forbidden, while the propagation of theoretically unclear cosmopolitan ideas of peace should be permitted. All thoughts and aspirations of this kind must be kept away from our fighting troops in particular. . .

It must appear particularly dangerous that this movement, which initially appeared more scientific, has recently sought to make contact with sharply internationally oriented socialist groups in all countries.

After all, such an appearance by people posing as representatives of the German intelligentsia abroad is likely to undermine the respect for German character and German efficiency that the people in arms have won for us.

(War Ministry, 7. 11. 15. No. 3740/15 Secret)

6

On the Western Front, the attacks of the French and English carried out with the greatest disregard for casualties, have indeed weakened our lines in some places, but the breakthrough, which was to be forced at all costs, has failed, like all previous attempts. Gentlemen, you can get an idea of the extent of the tremendous struggle if you consider that France alone deployed not very many fewer troops in Champagne than those with which Germany entered the war of 1870. There is no word, gentlemen, that would be deeply enough felt to pay the debt of gratitude of the Fatherland to our warriors, to our warriors who, despite an unprecedented enemy barrage, despite a multiple numerical inferiority, have opposed the enemy with their bodies a rampart that he has not been able to break through. Immortal honor to the memory of all who gave their lives there for their friends . . . The German people, unshakeable in their confidence in their strength, are invincible. It is to insult us if one wants to make us believe that we, who have gone from victory to victory, standing far in enemy territory, should be inferior to our enemies, who still dream of victory, in endurance, in tenacity, in inner moral strength. No, gentlemen, we will not be bent by words. We will resolutely continue to fight the battle desired by our enemies in order to accomplish what Germany's future demands of us.

(Speech by the Reich Chancellor, December 9, 1915)

7

Simply waiting on the defensive, which would be conceivable in itself, would not serve the purpose in the long run. From their superiority in men and material, the opponents have considerably more forces at their disposal than we do. With this procedure, the moment would have to come when the raw strength ratio would no longer leave Germany much hope. The ability of our allies to hold out is limited, but ours is not unlimited . . . In Flanders, north of the Loretto Height, the nature of the ground prevents far-reaching operations until mid-spring. According to local leaders, about thirty divisions will be required south of this. An offensive in the northern section would also require the same number. However, it is impossible for us to assemble this strength at one point on our front.

(General v. Falkenhayn, speech to Kaiser Wilhelm, Christmas 1915)

8

Chancellor:
. . . it means insulting us if one wants to make us believe that we . . . should be inferior in endurance, in tenacity, in inner moral strength.
General:
. . . the moment will come when . . . would not leave much hope.
Chancellor:
. . . German people invincible.
General:
. . . the ability to hold out is . . . is not unlimited

9

Biggest hit of the present!
GERMAN BE YOUR GREETING!

Artistically designed greeting plaques, legally protected, 15 x 11 cm, for pinning on doors, business premises, restaurants, etc., for sale for ten pfennigs.

10

To the German Army, Navy, and Protective Forces[12]:

Comrades! A year of hard fighting has come to an end. Wherever the enemy's superior numbers have stormed against our lines, they have been crushed by your loyalty and bravery. Wherever I asked you to strike, you have gloriously won the victory.

Today, we gratefully remember, above all, the brothers who joyfully gave their blood to win security for our loved ones at home and everlasting glory for the Fatherland.

What they began, we will complete with God's gracious help.

The enemies of West and East, North and South are still reaching out in furious rage for everything that makes our lives worth living. They have long since had to bury any hope of overcoming us in an honest fight. They believe they can only rely on the weight of their masses, on the starvation of our entire nation, and on the effects of their outrageous and insidious worldwide campaign of slander.

Their plans will not succeed. The spirit and the will that unshakably unites the army and the homeland will be their miserable undoing: The spirit of fulfilling one's duty to the Fatherland to the last breath and the will to win.

So, let us enter the New Year. Forward with God for the protection of the homeland and for Germany's greatness!

Grand Headquarters, December 31, 1915
Wilhelm

---

[12] Protective Forces or Schutztruppe. German colonial military in Africa from the late nineteenth century until 1918.

11

Do not draw too much attention to forthcoming successes, as this will diminish the joy of our victories or cause disappointment if they fail to materialize.

(Press briefing by the Chief Censor's Office of the War Press Office, August 20, 1915)

12

Diary entry by Sergeant Reisiger:

On the night of December 31, 1916, all German batteries, as far as I could hear, fired three volleys against the enemy at twelve o'clock sharp. The enemy did not reply.

I would like to know who gave the order. I have never experienced such a rage in the battery against this dog. This is an outright mess. There were tears in our eyes. After all, we are no longer children, and shooting is certainly our profession when we are at war. But the fact that we had to start the new year like this is mean. Why did we fire? Against whom, for God's sake? If it goes on like this, firing for the sheer pleasure of firing, without purpose, without "enemy," without (should I be ashamed to say "justification"?) ... simply because some stage stallion has come up with something particularly special as a New Year's greeting. Then sooner or later, things will come to an end that won't be fun for any of us. But I'm a soldier and have to keep my mouth shut.

Part II

Chapter I

1

Since the death of Jesus, it has been proven that one can die and not only has to die.

The most common question we hear today is: How much longer will the war last? Do not scold me lightly if I answer with a word that sounds like a joke, but which I mean very seriously: until it is over.

(*Berliner Lokalanzeiger*, April 21, 1916, Court Preacher Lic. Doehring)

2

God has visibly helped. He will continue to be with us.
(Nov 1, 1916; Wilhelm to Franz Josef)
God's help will lead us to a good end of our common struggle.
(Nov 1, 1916; Franz Josef to Wilhelm)
I pray that the Almighty God will give us his blessing.
(Jan 27, 1916; English Speech from the Throne)
Invoking the help of the Lord for your works.
(Jan 14, 1916; Order of the Day, Tzar Nicolaus)

Just as my blessed grandfather and I presented Ourselves as working under the highest care and commission of Our Lord and God, so I assume this of every honest Christian, whoever he may be. But whoever works with this attitude will realize that the cross is also an obligation! We should stick together in brotherly love...

(Kaiser Wilhelm, August 29, 1910)

... with Us ... with Us ... with Us ... with Us ... whoever he may be ...

3

February 16. Leave. After seventeen months, after five hundred days, Sergeant Reisiger was allowed to spend ten days in Germany. How longed for! How ardently, repeatedly, ever louder, ever more urgently desired!

The day had already been set twice. "Reisiger, next week—going to Germany—man, you're doing so well!"

Then the only thought had been, *Don't let yourself be shot in the head at the last hour!* Fear had set in.

Twice: vacation ban! They had ducked under this cursed word. Until Reisiger was finally on the vacation train.

And then? Then ten days, 240 hours passed.

Report: "Sergeant Reisiger back from home leave."

"Good, relieve trench observation immediately."

Reisiger sat in the bunker. It was raining, a greasy night under a greasy sky.

The day before yesterday in Germany, at the white-covered table between father and mother, at a loss for a word, a conversation. Yesterday on the train, among silent soldiers—now here.

Had he been on vacation? What did Reisiger know about it? It was nothing more than a movie, shot too quickly, clumsily edited, rushed, overheated, torn to pieces. Lines of catchwords, incoherent, unfounded, a jumble without order or law.

Title: Hunger!

Announcement:

For the current week, the following food will be distributed per head of the population: 10 pounds of potatoes, 210 grams of sugar, 100 grams of butter, 100 grams of lard or bacon, 80 grams of margarine or artificial cooking fat, 125 grams of legumes or cereals, ½ liter of milk daily, 25 grams of fine soap, 125 grams of other soap. Allocation of fresh meat is based on the total quantity available.

People stand in front of the stores. Now, in winter, early in the morning from 6:00 a.m., in two lines, the very old and the very young. From one leg to the other: "Yes, the war . . ."

"Well, our good people in the field will manage ..."

"I can't get by with eighty grams of margarine ..."

"And if I tell you that my son has written that they haven't seen a morsel of fat at the front for days now?"

"Of course, we still have it good and mustn't complain ..."

"My child is now six years old; do you know what it weighs: thirty-three pounds, believe it or not."

The shopkeeper pushes up the shutters: "There's only soap and sugar today. The other things are due in the afternoon. Well, you'll all have to be patient."

The people get the soap and sugar. They stand for another three hours in the early afternoon. In the evening, with trembling knees, they look gratefully and lovingly at the artificial cooking fat, cereals, frozen potatoes, and 100 grams of butter.

Confidential communication from the
High Command of the Marches[13] to all newspapers:

Publications about public disturbances that took place this evening in front of the Invalidenstraße market hall (Berlin) may not be made before the official investigation and corresponding notification to the press. (2. II. 1916) Further notifications about the incident in the market hall other than the notification given by WTB are not permitted.

(Oberkommando i. d. M. 3. II. 1916)

Wolff Telegraph Bureau WTB[14]:

In the market hall on Invalidenstraße and Ackerstraße, an iron stove was knocked over today during a rush of the public to buy lard; no person was injured. This incident gave rise to rumors of riots and the use of weapons by the police squad, but these are completely unfounded.

---

[13] High Command of the Marches or Oberkommando in den Marken. Prussian military authority for the suppression of public unrest. During the First World War, the commander-in-chief of the OKiM acted as military commander for Berlin and the province of Brandenburg.

[14] WTB: German press agency. One of the three main European telegraph news monopolies along with the British Reuters and the French Havas.

The new gardening textbook of the Peterseim-Erfurt flower nursery, suppliers to His Majesty, the German Kaiser, contains this and much more:

The health of a person is reflected in the chamber pot, the healthy or unhealthy agricultural condition of a people in the cesspool. The proud Roman Empire perished because of its cesspool economy. It is not war that destroys a nation, but only the state of the fields. It ultimately destroys a nation or makes it powerful. The number of marriages and children depends entirely on the price of grain.

The annual fecal matter of one person is enough to produce seven hundredweight of rye grain on one acre. Millions of quintals of bread grain worth of fecal matter are lost every year and discharged into the rivers through water flushing systems.

"The diligent hand shall rule, but the slack hand shall pay 'interest,'" Proverbs 12:24.

The film roll tears.

The white wall is so blinding that you have to close your eyes. But the motor hums, motor hums, motor hums. And then, on the white wall, enormously enlarged, distorted, gigantic, two hands, fingers, as thick as trees, groping and searching. The operator wants to mend. The motor whirs, the motor whirs, the fingers press and search and grasp. At last:

Title: The woman!

How was that? All those women on the streets, in the offices, in so many positions? What was that like?

There they are, woman next to woman, turning grenades. Leather pants, their breasts concealed in a leather jerkin; their hair, blonde perhaps, thin and flowing perhaps, concealed in a leather cap.

And the heart?

. . . I have spoken to many, many women: tame ones, whom I had hitherto thought limited, and they were clear and strong in their hatred; tender ones, who were only occupied with themselves, and hatred now made them forget themselves; childless ones, who had long been devoted, and who now complained: if only I had a son to avenge us on England; mothers who have no happiness in the world but through their sons: they gave them joyfully, and

they could no longer say clearly: for the Fatherland or out of hatred against England. (Ida Boy-Ed, on the day of the English declaration of war.)

Here they come. Engine drivers and chauffeurs, letter carriers and waiters, clerks and cashiers. Women, incomprehensible, of the female sex!

Of the female sex?

What was it like when you were on vacation?

Didn't an emotion break through from time to time? Could it not happen to you, Reisiger, that the eyes of the young woman in ragged men's pants, the coal pannier on her bowed shoulders, that these eyes suddenly flickered? That the mouth, dusty black, suddenly lifted its corners into a soft curve? That the hand, broad and dirty, trembled? That from the chest, gasping, the hills grew. That nothing was a lie anymore; that no war could disgrace a woman into a man?

What was it like, on vacation?

The many women, without a husband? War wedding in the afternoon, an embrace that shattered the body, at night, farewell, separation, in the first morning. And then, day after day and night after night and month after month—without a man?

The movie rolls: open arms. A pillow torn against her lap at night, cooling. And waiting. When will my husband come? Open arms. The body starved, exhausted, exposed, tossed back and forth between the sheets: A man—when will he come? Day after day and night after night: When will a man come?

Marriages broke up; hearts plunged into confusion—the man in uniform, every man in uniform: That's my man.

Letter from a warrior's wife, publication forbidden:

Since most of the women carry the result that the vacationers left behind and the same are already looking forward to the coming war boy, I do not want to stand back and be ridiculed. I also want to support patriotism. But I also don't want to go astray, but I want to practice German loyalty, because my husband has been in the field since the beginning of the war. Nature also demands its rights. I also hope that my plan will come to fruition, and the future will tell what our result will be. And our Kaiser needs soldiers. The request for leave must not take too long to be approved, otherwise the war will be over, and our plan will be thwarted to have a war boy.

Sincerely yours...

But did the man come? And if he was shot, lying in the field, torn to pieces? What else was at home, if not old people, cripples, children?

Bars, cages, walls. Behind them: captured enemies. It was dark; it smelled: Here is the man.

The municipal police office in Schwerin i. M. announces, April 1915:

Recently it has repeatedly happened that the civilian population has shown extraordinarily tactless behavior when prisoners of war are passing through. Not only have large crowds of curious onlookers gathered, but many spectators—especially the female part—have not refrained from showing compassion for the prisoners by crying, giving gifts, helping to carry luggage, and etc. The civilian population is advised that measures have been taken to prevent such behavior in the future under all circumstances.

Kreuzzeitung, April 1915:

The upholsterer's wife M. S. was sentenced to one month in prison by the Eichstädt district court for correspondence with prisoners. She had exchanged letters with French officers held prisoner in M.

And in the evening, despite the prohibitions, people lay in the frozen forest outside the town, in the abandoned barn, in the stable, their bodies pressed against each other, starving for a caress. A German woman and a French man, a German woman and an Englishman, a Russian, a Negro... For seconds, there was no war! No Fatherland! No German and English and French and Russian and Swahili! In seconds, the machine stopped: murder. In seconds, a German woman and a man whose language she didn't understand: only and nothing but woman and man.

Engine whirring, engine whirring, engine whirring—and the movie rolls off, rolls on, rolls—

And what else was there, Reisiger, on vacation...?

There were people in civilian clothes; they approached him confidentially, crowded the streets, followed us into the room, stared around the bed, laughed broadly, raised their fingers, threatened mischievously with their fists, rustled their newspapers, raised their glasses of beer in a toast.

The teacher from the upper school math class: "Well, Reisiger, are you making any progress? Seems to me you've gotten lazy up there at the front, haven't you? No victory for weeks, eh?"

The postmaster, retired officer from 1870/71: "Well, Herr Reisiger, back then, we, before Gravelotte, for sure, it was worse than now, now that men are hiding in trenches everywhere."

The pastor, with spittle on his lips: "Now dear Adolf, hearts to God, you remember, and fists to the enemy!"

And the many: "Now tell us a little about the war."

What was that, Adolf Reisiger? Tell me, a little bit, about the war. Started, after a long ordeal: "So, the enemy fired shrapnel . . ." And the story was already interrupted: What do they know about the enemy! What do they know about shooting? And "shrapnel," what do they know what a shrapnel is?

Stammering, silent: It makes no sense.

The father asks, "Well, boy, shall we have a glass of champagne today?"

His son, the Sergeant Adolf Reisiger, replies, "Yes, oh yes, I'd love to!" And thinks, *Was my battery allowed to fetch water tonight?*

Then Mother asks, "Do you want to eat a chicken today? Uncle Hermann sent one for you?"

The son replies, "Chicken, excellent." And thinks, *It will soon be lunchtime. Is there permission in the field today to open a tin of beef with the dried vegetables?*

Father: "And when there's peace, you can study as you please in Munich."

Reisiger thinks, *Will a man come home alive from the field as we prepare for a war of movement in the spring?*

His mother: "Oh my boy, who would have thought it? But you wanted to go into the field. We women probably don't understand anything about it."

Reisiger, to himself: *Now she's crying too. If only I was back at the front. If you women had forbidden your husbands and sons, forbidden them* with fervor and anger, to touch a single gun—but what do women know about war!

And the movie rolls and rolls and rolls. And, close-up, the Coupé III class, first between civilians, then, after twelve hours, only between military, comrades, and silent ones.

Drive across the Rhein. Swallowing in the throat: Here is the border. Germany on one side, and us on the other. Across, words, words, words. Only understanding no longer exists, no longer exists. You, over there, in Germany, behind us, don't understand anything about us. You can't understand anything about us!

The movie rolls slower, slower. The engine hitches, rattles, spits, and stops.

Reisiger sits in the bunker, in a greasy night, under a thick, slimy sky.

The enemy is silent. A few rifle shots hit the breastwork. Once, a voice cries out terribly. Reisiger sleeps, the telephone to his ears, listening as if in the twilight: All quiet in the West.

4

Reisiger's diary, April 2, 1916:

A Call

I have loved you, God—but where is your support?
Are you causing us all this pain? Do you turn us into animals?
Is it your work that our brothers bleed torn on stretchers?
Does your hand throw fire from the roaring maws of the guns?
Is it thy will that thy sons should be cut down innumerable like rotten trees?
Do you let guns bark wildly at our bodies?
Do you throw the flaming fires hot on our huts?
Do you allow our cities to be buried under the storm of your breath?
Lord! Great madness at you clutches me with a thousand arms!
I have loved you, God! Show mercy once more!
Lord! Once more be merciful! See the hands we raise in prayer!
Forgive us all our sins! And deliver us from life!

I wrote this poem a few days ago. I keep imagining what would happen if I were to read it out loud in front of everyone at the post office. But then they'd think I was crazy and lock me up. It's probably not worth it.

The only thing that would have to be said is that I'm beginning to think the war is the biggest mess there is.

I want to have the poem printed. I'll send it to editors in Germany. It's bad, but it says something. So maybe it will help.

Life is becoming more and more unbearable. Mosel shouldn't have fallen. And especially not in such a senseless way. Since he was a good soldier, and he was aware of it, he should have died in the front line. But an aerial bomb at the Protzen? Poor guy.

The new chief, Captain Siebert, is nothing like Mosel. He is undecided. Or unsure? Sometimes he doesn't know what to do. If only we still had Fricke. But his end was also foreseeable. You don't walk around the front line with impunity.

Where does life go? What happens next?

5

What we have experienced recently—the rejection of the Russian offensive and the battles at Verdun—are not, as our opponents pretend to believe, the outermost efforts of an exhausted nation giving its last, but the hammer blows of a strong, healthy and unconquerable people's army, equipped with human reserves and all the necessary resources. The attacks will be repeated until the others are worn down, and I promise the House that we will do everything in our power to achieve victory.

(*Minister of War Wild v. Hohenborn*, April 4, 1916, Reichstag)

6

Our newspapers should not write what the people like to hear and what, therefore, makes street sales more profitable. They should be the spiritual leaders of the people and write what is useful to the nation. And that is why they must not just work towards momentary effects but must look further ahead. Instead of trying to anticipate what our generals and troops will accomplish in the field based on a lack of expertise, they should rather try to calculate in advance what effect their publications will have at home and abroad. This is the right field of activity for the journalistic urge to see into the future.

(Press briefing of the Chief Censorship Office of the War Press Office, September 22, 1915)

7

... People like to hear ... Hammer blows ... with all aids ... insurmountable ... to anticipate ... until the others are worn out ... that is the right field of activity for the journalistic urge ...

8

F.A.R. 96 has been pulled out. The batteries are in some villages near Courtrai. Three weeks rest.

9

Court hearing in the Staff Quarters of I. Division F.A.R. 96.
 Present: Leader of 1/96 Major Klemper as chairman, Captain Siebert 1/96 as first assessor, Lieutenant Stiller 5/96 as second assessor.
 Secretary: Reisiger (promoted to Vize-Wachtmeister[15] three days ago)
 Accused: Gunner Rodnik, 1/96.
 Major Kl.: "Tell us what happened."
 Defendant is silent.

---

[15] Vize-Wachtmeister. Non-commissioned officer rank comparable to a staff sergeant (below a master sergeant).

Major: Kl.: "Well, is it true that you had sex with this woman, although you knew that the person was sexually ill?"

Defendant: "By your command, Major, no."

Major Kl.: "What do you mean, no? Did you not ... or did you not know?"

Defendant: "I don't know the girl, Major."

Captain S.: "Rodnik, if you keep lying, I'll lock you up immediately. Here, Herr Major, is another certificate from the battalion doctor that the man has severe gonorrhea."

Major Kl.: "Man, do you want to tell me that this came from peeing?"

Shouts. "You'll go to the penitentiary if you don't tell the truth. So where did you get the clap?"

Defendant: "From the military hospital ..."

Captain S.: "That's a lie. You were never in the military hospital?"

Defendant: "No, Herr Hauptmann ... but here in the village ..."

Captain S.: "Rodnik, don't talk nonsense. Now be reasonable. You should know me well enough to be able to talk to me. I don't want to be lied to, so now tell me. In your own interest."

Defendant: "I bought it ..."

Major Kl.: "Bought what? The girl?"

Defendant: "No, Herr Major. The clap. From an infantryman. But others did that too. He has gonorrhea ... and if you give him a mark, he gives you a little ... a little pus. And if you smear it on yourself ..."

Captain S.: "Others have that too? How many from the battery?"

Defendant: "When I was there, five more ..."

10

Suicides in 1914/15 in the field and occupation army; unconditional number: 621.

Suicides in 1915/16 in the field and occupation army; unconditional number: 1210.

(Seesselberg, *Stellungskrieg*, page 454)

11

On January 7, the Duisburg Regional Court sentenced the watchmaker Max Ranspieß to three months in prison for immorality within the meaning of § 184.1 of the German Criminal Code (distribution of indecent publications).

The accused, who is married but lives separately from his wife, had, as he was accused, used the current war period under the mask of a benefactor to go out on forbidden love adventures by carrying out the following maneuver: On September 19 of the previous year, he placed an advertisement in the "Duisburger Generalanzeiger" that read: "Young watchmaker takes on repairs on the spot free of charge for women whose husbands are in the field." Below this text was a number which, he hoped, would attract female reflectors. However, the advertisement had a completely unexpected effect for him, in which he had to answer for immorality.

The court was of the opinion that the defendant was only interested in being able to enter into illicit relationships with women whose husbands were in the field as a result of the advertisement. This was supported by the entire wording of the advertisement, namely the emphasis on the word "young watchmaker," then the "free" execution of the repairs, and finally the carrying out of the repairs "on the spot," i.e., in the women's homes.

The court did not believe the defendant's assertion to the contrary that he had no improper ulterior motives when placing the advertisement. The defendant's appeal was rejected by the Reichsgericht.

(*Leipziger Volkszeitung*, May 11, 1915)

12

The most interesting book of the present day!

The German woman and the prisoners of war

From the contents:
Iros: What attracted the woman to the prisoners of war?
Tlucher: Suggestion of the Foreign.
Dr. Sterzinger: Psychological analyses.
Anna Haushofer-Merk: Is the prisoner still the enemy?

Professor Henni Lehmann: We mothers! What I experienced, and how I made myself liable to prosecution.
Public prosecutor Laufer: Guilty?
Valentine Gebhard: The novel of the forester's daughter.
Ida Wagner: The working woman and the prisoners of war.
Ida Wagner: The woman as employer and the prisoners of war.
Valentine Gebhard: The beautiful Greeks of Görlitz, etc,. etc., etc.

The book openly and freely exposes the moral transgressions of the "prisoner lovers," seeks to fathom the causes, and drips healing balm on the wounds of German moral reputation.

Price stapled Marks 2.-, postage and packing 20 Pf. On request in sealed neutral envelope Marks 2,30

Every German man, every German woman should read this book. (Nürnberger Bücherei und Verlagsgesellschaft, Döllinger u. Co., Nürnberg 11, Königstorgraben. *General agents wanted everywhere*).

13

The relatives of prisoners should be careful in their choice of reading material that they send to prisoners in enemy territory; the inferiority of the reading material may indicate a low level of culture in our nation.

(Press briefing by the Chief Censorship Office, August 4, 1915)

14

F.A.R. 96 received the order for deployment one afternoon at the end of June. It was unusually clear for all batteries: Now it's off to the deepest mess. They had been given completely new guns; they were trained with new gas protection equipment and rubber masks that they wore like a muzzle. There were steel helmets.

A look at the army reports for the last few days . . . Where are we going? It was not until the last infantry standby quarters that the regimental orders gave precise instructions to the individual batteries as to where they were to march.

When Lieutenant Linnemann gave the order by telephone to the batteries, he added, "The regimental commander wishes to see the staff sergeants on officer's duty and all officers at the regimental staff headquarters this evening at eight o'clock for a beer evening."

A battalion order was attached for 1/96: First Lieutenant Rosstorf 2/96 and Lieutenant Stiller 5/96 were transferred to 1/96 with immediate effect.

The two gentlemen reported to Captain Siebert at about half-past eight o'clock in the evening. Reisiger was introduced to them. They rode together to the regiment.

Their quarters were in a magnificent French castle. Candles had been placed in all the windows. Long wooden boards had been laid out on the lawn in front of the building, and lanterns—red, green, and yellow—had been placed above them.

Very beautiful. The air was warm, and at night, the sky was infinitely soft and starry.

The beer evening was "informal." The staff sergeants of the individual batteries sat at a special table, but it was made clear to them right at the beginning that military forms were not desired today. Linnemann smiled. "Any one of you could find yourself in the embarrassing position of having to lead the battery in the next few days. You know..."

The colonel appeared like a civilian in long pants with patent leather shoes without a cap.

Music had been borrowed from a pioneer commando. Six men sat on the balcony of the castle with wind instruments and played. Waltzes, marches, later soldiers' songs, tasteless enough "Morgenrot..." and "Ich hatt' einen Kameraden..." Two melodies that didn't really fit in with the mood, which you don't like to hear in the field.

So what else could you do but drown any emerging sentimentality in beer? It was done thoroughly. After a very short time, there were officers at every table with their heads in their hands, stubbornly looking into their glasses and humming along to the music.

There were few of the staff sergeants who Reisiger knew. Most of them had only recently arrived; their predecessors, from his transport, were dead. Or wounded in Germany.

Drinking beer was a punishment for Reisiger. Here, it was done according to orders. It was enough for any officer, even the youngest lieutenant, to nod at you, and then you had to pour the booze down your throat. Reisiger's mood slipped more and more. He didn't understand, felt strange, had to laugh, and didn't really know why.

There was a clock in the castle tower. When it whined twelve, the colonel stood up and banged on the glass. He said words that no one would have expected him to say. Was the night to blame? He spoke very tautly, choppily, but you could tell that there was excitement behind his words: "Gentlemen! Officers of the 96th Regiment! You all know where we're going tomorrow. A soldier does not talk about what he is ordered to do. A soldier, an officer, does his duty. And, gentlemen, I know that even those of you who are not officers by profession will fulfill this duty. The enemy, gentlemen, is making the last effort. Nothing will help him. And when victory is ours, gentlemen, I expect to be able to say proudly and honestly on behalf of my regiment: We, all of you, have contributed to it. Let's not fool ourselves. We have had to bleed more in the last few months than any other field artillery regiment. Gentlemen, there are thirteen officers and over five hundred non-commissioned officers and gunners lying in front of the Loretto Heights. Let us remember that, gentlemen. We also want to die, but we want to be victorious . . . Thank you, gentlemen . . . good night."

15

Grand Headquarters *July 2, 1916*

Western Theater of War

The major Anglo-French mass attack, which had been prepared for many months with unlimited resources over seven days of the strongest artillery and gas attacks, began yesterday in a width of about forty kilometers on both sides of the Somme and the Ancre Creek. From Gommecourt to the area of La Boisselle, the enemy did not gain any significant advantages but suffered very heavy losses. On the other hand, he succeeded in penetrating the front lines of the two sections of divisions abutting the Somme at certain points, so that it was decided to withdraw these divisions from the completely destroyed front trenches to the blocking position between the first and second positions. As always in such cases, the material that had been firmly embedded in the front line and rendered useless was lost. . . . The enemy air service was very active.

16

After a march that lasted twenty-six and a half hours with a four-hour rest break, the First Battery 96 stood among the smoking rubble of some village ruins on a rainy morning.

Lieutenant Stiller was in command. Siebert and Rossdorf had stayed behind in the division to find out more about the final deployment.

The crews were so tired after the long march that they lay down behind the guns in the rain, pulled tents over themselves, and slept.

The horses remained harnessed. They had been given sacks of feed; they chewed the dry leaf hay, exhausted.

The village was deserted. Not a soul, not civilian, not military. There was no house that could have been used as a dwelling. No roof, no wall. Just stumps everywhere, staring into the rain. Gray, tattered, crushed. The village street was so rutted that the guns could only be moved if all the gunners gripped the spokes and the horses were beaten.

Was the village still under enemy fire?

The front could not be far away. The tremendous roar of the artillery battle could be heard. When the battery awoke from their exhausted sleep around noon, freezing despite the warmth because the rain had soaked them all to the skin, one thought automatically occurred to them: Is the fire up ahead near or far? And secondly, we don't care whether it's near or far, we have to go in!

They shivered and froze more. But then, thank God, very unanimously, all their worries were displaced by another, more important thought: hunger.

One by one, they crept toward the steaming goulash cannon. *What do we have? What could there be—dried vegetables and dried fish?*

Of all things! But the important thing is that it's hot.

We ate. The crews were grouped around their guns again, with the sergeants next to them. The canvas tents had now been stretched between the trees. It was quite cozy.

Reisiger stood with Hollert. Lieutenant Stiller, a stranger and unsure of himself, squatted alone beside his lad; he looked dully ahead of him.

After all, he wanted to demonstrate that he was still there. He gave the order to move the ammunition wagons, which until now had been standing close to the guns, forward by a few hundred meters. He looked up at the sky, which was gray and rainy. "Even if no planes come, do you know if the enemy won't shoot as far as here?"

Hollert was in command and in charge of the ammunition wagons. He mounted his horse, rode among the rubble, and finally found a shot-up brick kiln a few hundred meters further on, just outside the village. There, he led the four ammunition carts, the field kitchen, and a two-wheeled cart with the officers' luggage.

The ammunition wagons were already parked. When the field kitchen was halfway there, the catering sergeant made them stop. "Herr Leutnant, the tea is ready; can't we serve it right away?"

"Yes! Gun by gun!" The operator of the left wing gun hurried through the sodden clay ground toward the field kitchen with canteens and cooking utensils.

The next moment, a short, inconsequential bang, seconds later, a sharp shake in the air, seconds later, at the spot where the field kitchen with the gunners stood, a cloud of black smoke that shot abruptly over the treetops. Had they even heard a scream?

The men at the guns ducked their heads and pressed themselves to the ground. Reisiger wanted to say something, rushed forward, and almost collided with Lieutenant Stiller.

The cloud of smoke had cleared. You could see the field kitchen, crushed, one wheel flung meters to the side, shredded beyond recognition, seven soldiers, the catering sergeant, the two horses of the field kitchen, the two horses of the officer's baggage wagon.

The driver and gunners ran around in a hurry.

Lieutenant Stiller was stunned. "How is this possible?" Helpless.

But then he suddenly shouted in an exuberant voice: "Don't just stand here and gawk," and turned to Reisiger. "Reisiger, we have to find another place immediately, of course."

"By your command, Herr Leutnant . . . if I may suggest, the best place would be in the middle of the open field. After all, it is unlikely that the enemy has seen us. So, if he keeps shooting, he must mean the village."

"Battery mounted!" Led by Stiller, they galloped across the field toward the enemy. The guns were spread far apart. They stopped at irregular intervals in a cornfield. Rye had been sown. The stalks were very high; they even offered protection from airplanes if necessary. In addition, the horses could finally eat as much as they wanted.

On the right wing of the battery, spread out in exactly the same way, was the ammunition column.

The lieutenant stood apart from the sergeants.

The men had only one thought: damn it, now we don't have a field kitchen anymore.

Stiller: "Wachtmeister, what are we going to do with the dead?"

Burying the dead was difficult. You couldn't even tell who was lying there, unrecognizable! Hollert went from gun to gun using his battery list, and finally he had everything in order enough to put a cross behind some of the names. *We won't inform the relatives until the next few days*, he thought. There's no hurry anymore.

Then he had some drivers push everything into the blasting crater and pour earth on top.

Toward evening, the captain and Rossdorf arrived. "The battery will remain tensed under all circumstances. The marching order can come any hour."

17

Life awakens with the night. It has stopped raining. The velvet of the sky hangs incredibly low, once again studded with stars. The vault of the sky is divided into two parts. Light. And dark. Where the two parts meet, there lies waiting, 1/96.

The noses of the horses, and the gazes of the gunners go inward over the western part of the vault. From the zenith, it has lost its velvet, is first tinted slightly whitish, then takes on a more yellowish hue, falls into red, deeper into blood red, and finally into flames. There is the front. There it lies, a single incessant, never diminished, unstoppably steady thundering cloud. Without details. Without crescendo. Without rise and fall. Occasionally, of course, a powerful note wants to break out, to assert itself, swollen by its overwhelming strength. But the wave does not tolerate this. There is no single roar here. No single shot is valid here. No piston of a machine works separately here! The horizon toward the enemy is a single rolling wave of fire.

And the other part of the vault of the sky?

Velvet still above the waiting, blue gray sparsely interspersed with stars, and then quickly falling away into a deep black. The silhouette of the battered and skeletonized village stands jaggedly in between.

Incessant voices behind the battery, tired feet, the rumble of carts, horses' hooves gliding laboriously through clay. Reinforcements.

Can you sleep like that?

Most of them are awake, looking around, talking from time to time.

A tent has been built for the captain and Rossdorf. You can't hear anything from there.

Behind the Third Gun, a canvas tent is laid over the barrel, pegged to the ground on the right and left. Sergeants Hollert and Reisiger are lying in it.

Hollert is asleep. Reisiger has his eyes closed. But every time he is about to fall asleep, he startles up again. One time, his horse is standing with its front legs in the canvas; another time, he suddenly feels the animal's soft mouth on his shoulder. A little later, he hears a cry from somewhere.

It is already getting light in the sky. He still can't calm down. Then he hears someone calling softly for him. He jumps up and looks out of the tent opening. Lieutenant Stiller is standing there. Reisiger buttons up the collar of his coat, puts on his cap, and quietly crawls out into the open.

The lieutenant smokes a cigarette. "Reisiger, I'm sorry if I woke you. It's impossible for me to sleep." He adds apologetically: "It's unbearably hot in our tent . . . Why don't you stand at ease? Or, come on, let's walk up and down a bit . . ."

Reisiger looks the lieutenant in the face for the first time. *Younger than me,* he thinks. *Very young. Doesn't give the impression of being an active officer.*

"You're a student too, Reisiger."

"Jawohl! Herr Leutnant too?"

They walk slowly along the sleeping battery. There are only a few guards, pipes in their mouths, sleepy and freezing.

It turns out that the lieutenant is only a little younger than Reisiger. They experienced the mobilization in Munich at the same time.

"It's a pity we didn't get to know each other. Munich isn't really that big . . . We must have often sat in the same lecture hall . . . . At Kutscher's, wasn't it?"

In a matter of seconds, Reisiger's thoughts start to jumble. What nonsense! Drumfire—they call it drumfire now—and here they're talking about peace. And it's the end of June. And two years ago: "Then perhaps Herr Leutnant was even on the trip, back then, to Salzburg?"

Stiller grabs Reisiger by the hand. "But of course, dear. That's downright ridiculous. The open-air theater back then? And afterwards . . . that was nice."

Lieutenant Stiller and Sergeant Reisiger walk back and forth behind the battery and talk about peace, as if the student Stiller and his fellow student Reisiger were exchanging their experiences of the last days before the end of a semester.

But finally, Stiller waves them off, half laughing, half serious, almost officiously: "Inter arma tacent musae."

Reisiger reaches into his breast pocket and pulls out a postcard: "By chance, my history teacher, my former one, writes me a postcard. But he doesn't say 'tacent,' he says 'taceant.'"

They both laugh.

"It's a shitty existence," says Stiller and checks his watch. "Five o'clock. I'll have to wake the captain."

18

The battery has dug cooking holes and boiled tea, each gun for themselves. Everybody is sitting on the ground. Someone comes up with the idea of organizing a morning wash at the well that was discovered in the ruins of the village earlier. The captain agrees. So, some of the troops go into the village.

Suddenly, a dispatch rider appears. "Battery Commander First Battery?" It is a young lieutenant, the regiment's staff officer. Siebert reads the message he receives and shows it to Rossdorf. "Herr Oberleutnant, you lead the battery. I'll ride ahead, right through to the position. You follow until Sergeant Reisiger comes for you."

The battery sets off at a walk. Captain Siebert, Reisiger, and the telephone sergeant Kern ride to the front at a trot. The distance between the battery and this troop is soon so great that they can no longer see each other.

Siebert rides cross-country with his companions. A map is not necessary, at least for the time being. There is no better guide than the drumming at the front. And no mistake either. The path must be the right one: The drumming gets stronger and stronger. The monotonous sound breaks down more and more into individual explosions and into countless individual shots that roll incessantly one after the other.

The ruins of a village lie under heavy artillery fire. You can't get around them. Through! The three riders put in their spurs, the captain in front, the other two behind. Chest and head lie on the horse's neck.

There is life among the rubble. As the beat of the hooves hits the stones, many curious heads go up to the right and left of the devastated road: infantry, waiting infantry.

After the village an avenue, quite well preserved. Very tall poplars on both sides. Only rarely is a trunk frayed, branches lying across the road. Trotting, sometimes galloping, when the fountains of enemy shells come up too close out of the ground.

The horses speed up with necks stretched, nostrils flared, and deep red pits. The captain leads the way, his binoculars occasionally beating around his ears, followed by Reisiger, then Kern.

The avenue takes a sharp bend. Despite heavy artillery fire, which strikes from time to time and causes a few casualties with almost every shot, the infantry is once again waiting in the corner. Curious and gleeful, they look at the riders who are galloping around the corner.

Onwards! A hill on the left.

Siebert: "Up here, a kilometer at most, that will be the position!"

You don't quite understand what he's saying, but you can interpret it. Reisiger notices that the captain has a snow-white face and snow-white lips.

You let go of the reins. The horses make a few movements, then you are standing on an unlimited flat terrain.

A terrible sight!

There is hardly a square meter on this immense terrain that is not ransacked and torn apart by enemy fire. Drumfire!

The captain makes a movement with his raised left arm. It wants to indicate which direction to take. Reisiger pauses for a moment. A few meters in front of his horse's head, a black Vesuvius is already separating him from the captain and from Kern, who is bouncing past him.

The spurs in the horse's body, for God's sake, don't take them out, nothing but forward! Better an end with horror than this horror without end!

Gallop, gallop! You can't miss this field with the flaming black bushes growing up meter by meter. Follow the captain! He is fifty meters ahead. Aha, there's a willow up ahead. To the right of it is a small rise in the ground. That must be the position. Well, the horse is lame. Yes, dear Haybelly, that won't help. The captain rides as if he's being paid for it. If we get to the front alive, I'll eat a broom. *Golly, if the horse hadn't jumped aside now! Animals have instinct. That's what you call a barrage.*

All thoughts are blown away into nothingness by a crash and a bump that hits Reisiger in the chest. He disappears in a stinking white cloud, realizing at the same moment that a jolt goes through the horse. His head lurches forward, his forehead almost on the tips of his boots. His eyes go black for a moment; he feels a terrible weight on him and flaps his arms. The horse rolls on top of him, over his thighs. He wants to tear himself free. Another shot whizzes past his head so that he pushes himself close to the rolling horse. He tries again to free himself with his legs, gets under the animal's neck, jumps up, and kneels down. The horse thrashes its head. A lung hangs out of its chest, gasping and inflated. Reisiger realizes that the horse has been hit. Kneeling, he looks forward. Aha, that's the captain right at the pasture; the sergeant already dismounted too, the two horses on the reins.

His horse is lying on its back, its neck pushed unnaturally far upward, its gums pulled up, large yellow teeth bared, with a stream of blood coming from its chest and the corners of its mouth. "Poor Haybelly!"

Reisiger would like to lie down with his horse. He has an intense need to do nothing but cry. But when he realizes that tears are running down his cheeks, he pulls himself up, takes the pistol, and places it behind the animal's ear. Pulls the trigger. He sees the horse fall peacefully on its side. Runs forward.

It takes him half an hour to cover the distance that takes two minutes in peace. At last, he appears, the left side of his uniform soaked with his horse's blood, his face and hands filthy, at the captain's side.

My God, what a position! How can a battery be brought up here? How can it fire?

But there is nothing to think about.

"Take our horses," shouts Siebert, "let them run for all I care, but get the battery!"

Dear God, how are you supposed to get on a horse now?

The two animals are standing close together, heads between their legs, near a willow, trying to cover themselves. Reisiger rises, falls to his knees. "Your orders, Herr Hauptmann, battery get into position here immediately."

He crawls up to the horses. Then he gives himself a jolt, stands bolt upright, mounts Siebert's horse, takes the other one by the bridle, and gallops off.

The same way. The same fire. Only one thing can help here: Don't command the horse, trust it. Leave the reins loose. The animal knows more than the human.

So just one hand on the halter, curb bit loose, the other hand clinging to the saddle button, off you go at a gallop!

There lies the poor Haybelly. There lies an infantryman without legs. The horse jumps to the side, and the second shies away and wants to stand still. Finally, they both gallop on.

There lies another infantryman. There lies a group of seven men. There lies a pioneer officer. You can see the regimental number on his armpits. There lies an infantryman. There's a gun and its crew, including horses, including a lieutenant. Gallop, gallop, gallop! Without a second thought! A shot for every square meter? Perhaps not. But perhaps ten shots per square meter in the same amount of time. And you just have to see how you wipe through it. Now comes the hill with the ravine. Jumped into it.

Now comes the path, the avenue with the poplars.

There, the infantry.

Now the bend.

Further along the avenue, galloping, always the second horse behind you, beside you, in front of you—an excellent indicator of the proximity of the impact.

Still the avenue.

Now it's downhill.

The open field: There comes the battery.

Riding up to Lieutenant Rossdorf at a gallop, no. At a trot, no. At a walk, from the left. "Reporting for duty."

19

Captain Siebert crouches with Sergeant Kern behind the small slope to the right of the willow. Kern has unstrapped his field spade and is digging a hole. The turf is tough and difficult to pull out, but thank God the soil is not chalky, it is fat clay, from which the sharp spade cuts out one slice after another.

Siebert can't stand to watch. The fire that surrounds them from all sides urges him to get busy. He snatches the spade from Kern's hand and continues digging himself.

After half an hour's work, the hole is so deep that two people can kneel in it, pressed close together.

The two kneel as if in a prayer chair. Their stomachs pushed against the front wall; their elbows pulled up against their bodies; their heads sunk deep into the hole. So low that a shot, even shrapnel, must pass over them.

Moving in this situation is difficult. They sit still and silent. The scissor scope lies to the side of the cover, still in its leather case. Who will fetch it?

They feel as if they have been absorbed into the ground, as if they are stuck. But it doesn't help. Siebert tries to fish for the sheath. He finally crawls and pulls it toward him.

The telescope is placed between them. The glasses are centimeters high above the edge of the wall.

Siebert looks through it. He mumbles something, takes his eyes away again, and says, "It's completely nonsensical, you can't see anything. Just fire and smoke. No one will be alive anyway. I don't understand who, for God's sake, gave the order to drive up here in broad daylight."

"It's also questionable, Herr Hauptmann, whether a gun will come forward at all."

Siebert laughs. "So that means you and I are supposed to replace the whole battery..."

Kern looks around: "It already does not matter any longer. We can no longer go forwards or backwards..."

Siebert tries another look through the scissor scope. "If only we knew how far ahead of us our infantry is. Otherwise, we can't shoot. We'll only cause mischief in our own trench."

He lifts himself slightly from his knees and looks over the edge of the embankment. He has the feeling that the fire in the front area is slackening a little. The majority of the shots are now a good 100 to 150 meters in front of them.

It is worse behind them. All visibility is blocked by a thick white cloud with black streaks running through it. An eerie sight: The sun is just bursting through the wall, cutting through the white and making the black even blacker.

They turn around, marveling at the threatening swathes. And then a tremendous shock strikes their blood . . . The cloud wall suddenly tears wide apart in four places, gaping open with steam.

Siebert jumps up and gets back on his knees. "Hell, here comes our battery!"

An earth-shattering horse race has started. The arena opens, a modernized, 1916-style Roman combat game, advancing toward the ordered target with a force that only death can break.

The wall gapes open, closes, and the curtains are lowered again. The chariots are parked in the arena. The signal seems to be given for the race to begin.

There are only two spectators in the entire arena: Captain Siebert, Sergeant Kern. Their eyes dart back and forth from one carriage to the next. Bodies bent forward. Arms gesticulating. Thoughts shooting together: *Who will win the race? Who will win? Who will stay in place?*

First glance: Horses. On four lanes of the arena, black-and-brown patches are thrown up, press themselves to the ground, move off the track, turn, collide, break away, and fly in front of each other again.

Four carriages at the same height.

To their right and left, huge gates rise up to the sky, again and again, black columns and flaming candelabras. The carriages race forward through these gates.

The two spectators don't say a word.

The black and brown dots come closer, closer, closer. Now behind them the four battle wagons. Four guns of the Prussian 96th Field Artillery Regiment, the Protzen hard behind the horses, slung, staggering, and unsteady the four mounts. And more clearly now the leaders, out to the right or left with their bodies glued to the horses' necks and backs.

The two spectators lean backward against their lounge seats for a moment. One says to the other: "Unteroffizier, I can't believe it." The other does not answer. He is just thinking: *That there are people riding along . . .*

Then both: *That there are men riding along! Our men.*

Siebert is looking for Rossdorf. Aha, he's in front, in front of everyone. And, yes, between the Third and Fourth Guns, Lieutenant Stiller and Sergeant Reisiger.

Then his hand suddenly grips Kern's knee. In an unnaturally high voice, he shouts, "Over!" Kern sees a black cloud billow up between the front horses of the left wing gun. In one instant, the animals topple over each other. The next two into them. The two pole horses behind them try to get out of the way; too late, they, too, collapse over the black lump.

The gunners jump from the gun.

You can see them pulling the sidearms and writhing in the ropes of the harnesses. You see the gunner, dismounted. He plunges into the mass of horses and pulls people out, the drivers. He puts the first one aside. After a while, the second one too. They both remain motionless.

Then the pole horses jump up. With them, still in the saddle, the third driver. And the two horses are galloping again.

Siebert and Kern are now standing upright. The race is over. Three guns are here! Shortly before the position, they drop into trot and turn around. While still on the move, the gunners jump between the Protzen and the gun carriage. The gun carriages are unlimbered! The ammunition has been torn out of the ammunition boxes! The fourth carriage has arrived! After seconds, the guns are properly positioned behind the hill.

Back to the Protzen! Order: Take them back to the bend in the approach road and rest! Ammunition and rations must reach the position in the evening!

Chapter II

1

After a few minutes, no one in the battery is interested in the past hours. There is no reflection. How is it possible that we are still alive after this chase through a thousand deaths? There is no more trembling in the nerves, no more gnashing of teeth.

Instead, there is invigorating curiosity. What is going on here?

The enemy fire now lies in a huge barrage in front at the height, in the field of vision. So, presumably, there is the infantry. Very nice, you are immediately oriented about the position. From time to time, man-sized black shapes emerge from the clouds of smoke. It's familiar: This used to be a forest. The battle here will probably revolve around it. Whoever has the height controls the area. The task of the infantry must, therefore, be to hold out up there.

This is considered and discussed. At the same time, however, it is doubtful that there will be many alive in the fire up there, even with deep dugouts.

After a short time, even such concerns are dismissed. There is work to be done. The most important: to hide the position from the view of airplanes as quickly as possible! "Get some branches and tie them to wheels and barrels. You can go ahead and make Whitsun oxen out of the shotguns," says Lieutenant Rossdorf.

That's what happens. After a short time, a small wood has grown behind the hill.

Second job: digging ammunition holes! That happens too. A pit next to each gun, two meters deep, so that the baskets are really covered.

Third job: shelters! But what are you supposed to build shelters out of in the open that you can reasonably rely on? The willow on the left wing is the only tree in the whole area. It is out of the question. It is crooked; the trunk is weathered. So, there is nothing left but to dig holes for the people too. The captain arranges one directly behind each gun; three gunners remain there in any case. Everything else is allowed to stay in a few narrow trenches a few meters to the right of the battery position.

After the work, which takes several hours, it is time to eat. Field rusks are distributed, a quarter portion per man.

Now you lie next to the guns and let the sun roast you.

The fire on the heights makes no impression at all and is ignored. When now and then a few rounds whiz over the position, no one moves or raises a face. We are used to more difficult things from this morning. "That we've only lost four horses and the two drivers . . . my dear, we haven't had such luck for a long time!"

Captain Siebert sits behind the willow with Rossdorf and Stiller. They are also munching on their rusks.

Siebert: "Gentlemen, I think it's absolutely essential that we make contact with the infantry. Otherwise, we'll never get a clear view of the terrain."

The three officers look ahead. Rossdorf shrugs his shoulders: "We won't be able to establish a telephone connection."

Siebert: "Then we'll have to have signaling devices. The infantry does not have it any other way. And I won't take the responsibility of firing a single shot without knowing exactly where our forward position is."

Stiller tucks up his open collar: "If I may, Herr Hauptmann, I would go myself."

"Who are you taking with you?"

"I suggest Sergeant Reisiger. He's been on telephone duty long enough. If Herr Hauptmann says so, we'll both go into the trench first, and then, when we're back here, we can discuss with Herr Hauptmann where a permanent observation post with a telephone can be set up . . . . Reisiger!"

2

Stiller and Reisiger wander around in the foothills. They are looking for some kind of approach trench. There is nothing to be found. In some places, they see meter-wide whitish paths that suddenly start somewhere and run flat in the direction of the enemy. It is not clear whether these were ever trenches.

The enemy fire lies before them, incredibly concentrated, on the edge of the heights. So sharply defined that hardly a single shot jumps out of the compact mass of countless impacts. A thick black wall, impenetrable to the eye.

They finally give up trying to find a trench. Stiller looks around for the battery, then looks at the fire again. "Yes, it is no use, if we don't go in, we won't get to the front. So, what do you think, Reisiger?"

Nothing to say. Reisiger says matter-of-factly, "Your orders, Herr Leutnant."

Stiller: "Do we need the gas mask?"

He answers himself: "Well, it's not going to be that bad."

But then they both take the mask out of the canister and hang it in front of their chests.

They are closer to the fire. The roar is so loud that they can only communicate by shouting.

"It's enough," the lieutenant shouts, "if we get hold of a single infantryman. They must be somewhere."

He makes a dash. Reisiger follows behind. Both are lying in a shell crater. Two shells hack into the ground behind them. As the dirt splashes up, they take a running start and jump on. They fall into a shell crater again. Impacts to the right and left of them.

"Very practical, the invention of craters," the lieutenant shouts, "if you fall here, you always take cover straight away."

He is about to grin when a shot hits the edge in front of him. A good ten shovels of earth fly over their backs. They jump up, run bent over, lie on their stomachs, and jump up again bent over as it screeches above them. They both sit up to their knees in a large hole filled with brown water.

Stiller wants to be the first to crawl up the greasy belt to continue forward. He slips off and sits in the mud up to his navel. He must have stepped on something: Reisiger sees something puffing up out of the water, bloated—a body, an arm, a slender hand missing its thumb.

Without paying attention to the lieutenant, Reisiger leaps away in a single bound. A push immediately hurls him back to earth, into a hole next to it. Stiller pushes forward out of the smoke behind him.

They crouch together. Crawling further is impossible now. The crater is so deep that they could both stand upright in it without sticking their heads out. But they are crouching on the ground, not looking up. From the vibrations and the lumps of clay pattering at their feet, they can sense that fire is coming in all around them.

After minutes, Stiller notices a gap in the smoke in front of him.

He jumps up, through it; Reisiger follows.

The next crater is more than a jump away. They push forward on their stomachs. At last, they are at its edge. They see two steel helmets inside: infantry!

Yes, two infantrymen from the front line. Black faces. Uniforms smeared beyond recognition. Black hands. A rifle between their legs. A pile of hand grenades around them . . . . Both flinch as Stiller and Reisiger jump toward them. Then they recognize the officer; they raise their heads slightly.

Communication is difficult.

Even if you shout in your neighbor's ear, all words are shredded.

But what is there to talk about? The infantrymen don't seem to realize that you even have to.

"Where is your position?" One of them looks at Stiller with half an eye, and points his thumb at the dirt below him. "Here, Herr Leutnant."

"What? You must have a trench," shouts Stiller.

The infantryman shakes his head. "No, Herr Leutnant, we used to have one."

Stiller: "Who's in front here?"

The second infantryman: "Four companies. As far as the forest. But there won't be many more of them left today."

Stiller looks at Reisiger. The information is not enough.

"We have to keep going!" Then, once again, to the infantrymen, "Where is your company commander?"

They point to the left. "There somewhere. That's all we know."

"So let's go! Let's go!"

Now there are one or two infantrymen in roughly every shell crater. That's a bit of a reassurance. But you jump into a lot of holes where the crew is dead. You come across wounded. It doesn't help, go on! It's getting dark! But the battery must fire before evening! Because of the muzzle flash, otherwise, it is immediately compromised. Jump up. Jump up.

A crater, a hole in front of it: Here is a shelter! Stiller and Reisiger are pushed directly into the entrance by an explosion. They sense that they have steps underneath them and grope backward down them. "Infantry!" shouts Stiller. They get an answer, a questioning "Well?"

A young infantry officer is sitting by a candle. He shakes Stiller's hand, nods to Reisiger, and pushes two hand grenade boxes against the wall with his leg. "Please take a seat."

He explains the situation with the help of a map. The ridge, that is, everything about thirty meters in front of this dugout, is occupied by their own infantry, who hold out in the main blast holes. Not much more can be said; one can hardly gain an overview. He has been here for twenty-four hours and only knows that his company has lost about half its men so far. "God only knows how this will continue, but it's good to see that the artillery is making an appearance. A small consolation at last. Just shoot hard tonight! By the way . . . yes, comrade, I can assure you that the enemy was still a hundred meters behind the heights this afternoon. How is it now? I believe that as long as his fire is as it is now, nothing can happen to us. If the enemy puts it more toward you, down there . . ." He shrugged his shoulders and continued, "But they can come, we will give them a warm welcome."

After this conversation, it is clear to Stiller that the battery can shoot up to 150 meters behind the heights without endangering the infantry. "We will set ourselves up according to the map so that we can immediately lay down a barrage directly on the ridge if you request it . . . Agreed?"

"Very well!" The infantryman stands up and looks up at the stairs. "I hope you get back to the battery safely," he says. "Yes . . . I can't send you any messages. Telephone is useless too. But as you know, four red flares mean attack. I'd like to offer you a cigarette . . ."

He takes a crumpled white sausage out of his coat pocket: "That's all."

Farewell. The two of them crawl upstairs. Halfway up the stairs, a flame bursts into the darkness. The steps rise; there is a rumble. They lie waiting. After a short time, they push on. To the battery as quickly as possible!

3

I believe that we in the III Army Corps, as well as in the entire army, know that there can only be one voice about it, that we would rather leave our entire eighteen army corps and forty-two million inhabitants on the battlefield than cede a single stone of what my father and Prince Friedrich Karl have gained. In this spirit, I raise my glass . . .

(Wilhelm II, August 16, 1888, Frankfurt/Oder)

4

When 1/96 had been surprised by the sudden appearance of six L.M.K. ammunition wagons at about ten o'clock in the evening, the enemy fire had again reached the rear area after a long pause. The L.M.K. vehicles had not been persuaded to drive up to the position, despite orders to do so, despite shouting and swearing. During the approach, the crews had torn the baskets out of the wagons and scattered them haphazardly across the terrain. Then the wagons had scattered and disappeared.

The battery crouched in its holes. Captain Siebert was at the willow on the left wing, the Second Gun was manned by Reisiger, the Third by Rossdorf, and the Fourth by Stiller.

The enemy drummed in front and behind the battery.

A thunderstorm, incessant lightning. The fire burst several meters high with red-and-yellowish flames against the sky with every impact. A bluish full moon hung from it. It was fat and ponderous and had a pale, eerie light. All the men had whitish-blue faces and looked like madness.

Between the Second and Third Guns, a cannoneer lay exposed behind the hill as a flare post. He had his forehead and steel helmet pressed almost to the ground, leaving only a narrow slit open.

Suddenly, he screamed, struck backward, rolled down the slope, and came to a halt.

The scream had been heard; the Second Gun provided a new man as relief, who lay down in the same place.

Shortly before eleven o'clock, the moon was vertical over the battery, he shouted, "Barrage!"

The battery was prepared, the guns target fire was set at a distance of 1,700 meters. The alarm call had barely been shouted out when the first salvo was fired at the enemy.

Now everything was forgotten, the crouching in the hole, the ducking from enemy shells; officers, non-commissioned officers, and gunners operated the guns. The upright flashes of enemy fire were now pierced by the horizontal bursts of fire from the 1/96.

The captain gave commands, which were relayed very matter-of-factly. With anger, with hatred at having to lie on his stomach in front of something so senseless even now.

When Siebert then, with more red flares going up, redeemed the slow barrage to rapid fire, the crews glowed.

Basket after basket was torn out of the holes and flew empty to the rear.

That ate up ammunition. It was clear that the only way to continue firing like this was to get to the piles of spare ammunition that the column had scattered behind the battery.

Siebert intervened: Two gunners from each gun had to be freed to fetch the ammunition. Instead, the officers sat down at the gunner's seat. Reisiger jumped to the last gun.

But how could he summon the strength to tear himself away from the cover of the hill? The eight gunners were stuck on top of each other next to the left wing gun. Nobody made the first move.

Finally, the captain jumped up, jabbed one of them in the back with his foot, bent down to him, his hands at his mouth, and shouted, "Lümmels, get your ammunition now!"

The eight gunners pulled themselves up, straight for a moment, then bent at the knees, their chins on their chests. Shots from the enemy lay close in front of the battery. Shrapnel whistled over the hill.

Seconds. The gunner Wehrstedt jumped backward, lay down on his stomach after the jump, and crawled on all fours. He was three meters behind the gun when the second one followed. The third. The rest. The eight formed a chain. Wehrstedt lay in front of the largest of the three ammunition piles. He grabbed a basket with his hand, dragged it down, pushed it, and hurled it to the man behind. He passed it on. He went on. The ammunition supply for the battery was running.

Rapid fire!

The mountain was almost empty.

Then suddenly, the crews at the guns pulled their faces back: A broad blue flame shot against the sky, higher than a three-story house. Black chunks flew between the fire. Five of the chain of eight gunners leapt to their feet and ran for the battery.

A deep roar rose above all the crashing: One of the ammunition mounds had taken a direct hit. Wehrstedt and the two men behind him were torn to shreds.

Ammunition! How far does the ammunition reach?

Stiller was kneeling at the Third Gun; he jumped to the second, to the first, to the captain: "Herr Hauptmann, if we are to continue firing, we absolutely must try to get the rest of the ammunition here as well."

There, new red flares on the heights!

"Rapid fire!" shouts Siebert. The call faintly reaches the Second Gun, where Rossdorf waves the Third and Fourth with raised fists; the rapid fire rolls again.

Then the captain turns to Stiller. "Take the lead for a new relay yourself. Two men from each gun. Go on!"

An impact in the willow. It splits in two; one half remains standing, burning at the tip, and the other strikes black forward.

Stiller pushes two gunners of the First Gun between the ribs. "Come along," jumps with them to the second.

It does not go any further.

The battery is suddenly bathed in a bright white light, a burning cold brightness.

Everyone looks upward.

Above the position, perhaps ten meters high, hover two large yellowish parachutes with magnesium flares attached to them. The next moment, an airplane appears between the flares. A bird of prey of unprecedented dimensions, it swoops over the battery, makes a sharp turn, and swoops down to six, seven meters.

His machine gun splashes between the gunners.

You can see every bullet. He shoots with tracer ammunition. Small yellowish wasps, glowing with tiny spraying tails, hiss around.

Everything is on its belly. Somewhere, someone is shouting "cover." But how can there be cover here against the enemy from the air? And then a man at the Second Gun leaps up, his body bent upward and claps down again. At the Fourth, another is already shouting, kneeling, pressing his hands in front of his stomach. Is there rescue? One looks up. The flares go out. The plane disappears.

No one had been prepared for this. Air raid at night, down to five meters, with M.G. *That was new!*

The battery is paralyzed.

Finally, Siebert checks the position. There, at the Second Gun, next to Reisiger, lies the gunner Kolpe, who had jumped up earlier. His face is still turned to the ground. Siebert quietly lifts his head. Blood is dripping from the crown of his head. "No more use," Siebert says to Reisiger.

And to the Fourth Gun.

The gunner Seelow is still writhing there, his arms pressed to his stomach, lying backward with his knees raised. Around him squat the comrades of the gun, the medical sergeant, Rossdorf.

But what is to be done? They have torn open his coat and trousers: shot in the stomach.

The medical sergeant is helpless. How do you dress up something like that? He pushes cotton into the bullet hole, and it comes out again immediately.

Siebert wants to give some advice.

Then, as if a vortex was sweeping up the position, it whooshes in the air again. It's light again, spraying again. The plane has returned!

Everything is crowded together, pushed close to the bottom of the slope.

As Reisiger lifts his head for a moment, he sees three yellow spheres shooting upward from the plane.

He immediately puts his head back down.

Seconds later, the crews are flying around in confusion: four hits from an enemy battery are sitting directly on the cover!

The four shots are followed a few moments later by four more, three at the same spot, the fourth a little in front.

Where is the captain? Reisiger jumps to the Third Gun, and pushes his way to the Fourth. "Herr Hauptmann, the airman has uncovered us! Yellow flares! Now the enemy is on to us."

Siebert wants to answer. There's another crash. Four columns of smoke, filled with fire. Four shots directly behind the position.

Siebert shouts, "Take cover!" He simultaneously jumps out to the right into one of the larger holes. The gunners of the Fourth Gun follow behind.

Blue flames: Two hits on the hill, two against the stump of the willow.

"Everything into the holes!"

The enemy drums on the position of 1/96.

5

The 7.5 centimeter shells of the light field artillery, which weigh 5.6 kilograms and have an explosive charge of 0.608 kilograms, penetrate 1.80 meters into the ground, 12 centimeters into concrete, have a total impact and explosion force of 230 meters, and hurl 508 splinters around. The penetration depth of an impacting 15-centimeter shell into earth is 4.10 meters, in concrete 39 centimeters, the explosive charge weighs 4.86 kilograms, the force of the explosive charge is 1900 meters, and the number of fragments is 2030.

A 30.5 centimeter shell has a weight of 324 kilograms, develops an explosive force comparable to a D-train of 10/50 ton wagons at 85 kilometers per hour, hurls 8110 splinters around and penetrates 8.80 meters deep into earth and 90 centimeters into concrete.

(Friedrich Seesselberg, *Der Stellungskrieg*, page 260)

6

Sergeant Reisiger and Gunner Winkelmann are sitting in the hole behind the left wing gun of 1/96.

The enemy is drumming.

The hole is so wide that the two can sit opposite each other. They have their legs drawn up against their bodies, their knees pushed into each other, their elbows resting on them. Their hands hold their faces. They look down.

The enemy is drumming.

The 7.5-centimeter shells of the light field artillery penetrate 1.80 meters into the ground.

Occasionally, one of them stirs, twitches his knee, and pulls his chin closer to his chest. A sign for the other: I'm still alive—and what are you doing?

The enemy is drumming.

The living force of a 15-centimeter shell is equal to the momentum of two marching infantry divisions.

After a while, the other one. He can't speak. His neighbor wouldn't understand him after all. So, he twitches his knees. Or pulls his chin tighter to his chest. That means: *I'm still alive . . . so how are you?*

The enemy is drumming.

A 30.5-centimeter shell hurls 8110 splinters around.

Your head is only raised, jerked up at a lightning-fast moment, when a column of fire hisses up so close to the edge of the hole that you can feel the embers. Then you lower your eyelids, and a shot is missed.

The enemy is drumming.

There are 7.5-centimeter shells, 15-centimeter shells, 30.5-centimeter shells.

Sometimes, the two bodies of the men in the hole are thrown hard against the wall. The ground has risen violently, buckling under the impact. The impact is felt by the men. The shoulder crashes against the clay, against the steel helmet.

The enemy is drumming.

Reisiger shoots forward, and his face smashes hard against the edge of Winkelmann's steel helmet. His gums are bleeding. He spits and straightens up again.

The enemy is drumming.

Both give up their crouching position at lightning speed. There is a competition to see who is on top and who is on the bottom as they smear their backs along the clay wall, their noses pressed deep onto the bottom of the hole. This happens when a beam of fire is directly above them.

The enemy is drumming.

Once it hits hard against Reisiger's paddock lock, he feels the blow painfully on his stomach. Something rolls between his feet. He grabs it; it is a red-hot splinter.

The enemy is drumming.

The two men stare down. The light becomes dimmer. Reisiger sees Winkelmann's boots, Winkelmann sees Reisiger's dirty gaiters. Reisiger raises one palm upward and stares at it. Winkelmann undoes the hooks on his collar and two buttons on his uniform coat.

The enemy is drumming.

Hungry? No. Thirsty? No. Smoking? Reisiger reaches into the pocket containing the bandages and pulls out two cigarettes. Winkelmann gets one. He tries to light a match. It goes out. The same two or three times. The two of them smoke. They tear the smoke into their lungs, and they blow it between their knees to the ground.

The enemy is drumming.

Reisiger looks Winkelmann in the eye, smiles, points to the cigarette, nods. Winkelmann smiles too, nods again.

The enemy is drumming.

It begins to rain. The rain is as thick as fog. The two take their cigarettes into the hole between their hands so that they don't get wet.

Fire at the edge of the hole. The two of them slide to the ground and have to stick their cigarettes in the mud as they fall. Out.

The enemy is drumming.

The rain gets heavier. No more fog; thick threads; it is pattering on the steel helmets. They move closer together. Reisiger's knee under Winkelmann's chin. Winkelmann's knee against Reisiger's chest. The rain is a torrent, pouring along the steel helmet, dousing the curved backs, soaking between the edges of the collar, and pouring into the boots.

The enemy is drumming.

The clay floor is as disgusting as artificial honey. The men in the hole can no longer move normally; all their limbs slide back and forth on slippery sauce. It is almost impossible to sit. Just one movement of the head is enough to throw the body off balance. You brace yourself with both hands on the ground to support yourself.

The enemy is drumming.

The rain pours.

The water in the hole rises slowly.

Reisiger reaches into his coat pocket and pulls out a notebook; it is already half wet, and he pushes it through his coat and shirt against his chest. Winkelmann wants to save the bandages. He pulls it out; it is thick and greasy like a used sponge. He lets it fall to the floor. And they both see it floating, a small sinking ship.

The enemy is drumming.

The rain pours down.

They are soaked to the skin. The water in the hole has risen so high that it washes around the elbows of their outstretched arms.

The enemy is drumming.

How late?

When the plane came the second time, it had been around twelve o'clock at night. Reisiger pulls a hand out of the water and pulls up his sleeve; his wristwatch is still working. He shows it to Winkelmann: seven o'clock in the morning.

The enemy has been drumming for seven hours.

The 7.5-centimeter shell hurls 508 fragments around, the 15-centimeter shell 2030, the 30.5-centimeter shell 8110. Penetration depths into the ground 1.80 meters, 4.10 meters, 8.80 meters.

The enemy is drumming.

The rain pours.

Hungry? Reisiger tilts against Winkelmann: "Do you have anything to eat?" Winkelmann pulls a hand out of the mud, rummages behind him in his coat pocket, and pulls out a black canvas bag. He doesn't need to open it. As he touches it, a thick yellow broth flows out. They were field rusks. Reisiger nods. The greasy bag sinks into the mud.

The enemy is drumming.

Tired?

Terribly tired. It would be best to let your head hang ten centimeters lower so that your snout is stuck in the thick sludge. Then you could sleep.

But it's too cold for that.

They are sitting in water almost up to their shoulders.

The enemy is drumming.

You know that there is no rescue. Unless . . . the rain stops or the enemy calms down.

The enemy is drumming.

The rain pours.

How late? Reisiger pulls his hand out of the dirt and the mud creeps up to his elbow, then plods down. The sleeve is difficult to roll up. The watch glass is muddy. I wipe it off with my chin.

The watch stops.

The enemy is drumming.

There is a hole behind each gun in Battery 1/96. There are two or three men in each one. There are four more ammunition holes, probably still half full. There will also be two or three men in each. There are also two slightly larger holes on the right wing. The rest of the battery sits in them.

The enemy is drumming.

The rain flows.

The water reaches Reisiger and Winkelmann's chins. It is impossible to remain seated. They have to stand up. They drag their limbs out of the mud. The whole body has already become a part of the soil. It takes an effort to free it. They both stand, knees together, backs free, arms folded across their chests, hunched over. The steel helmets lie edge to edge, forming the only support.

The enemy is drumming.

Sometimes, you have to sink to your knees again, for seconds at least. Then you can see that the water level is making waves. The bottom churns up so that it almost tears your legs out from under you...

Then, the rain stops, as if blown away.

Is the enemy still drumming?

When the walls stop shaking for a moment, Reisiger lifts himself up, rests his hands on Winkelmann's shoulder, and looks over the edge of the hole.

It is unimaginable: Life lies before him. There is the sky, visible right to the edge of the horizon. From some of the holes, he can see the tops of the steel helmets. It moves mysteriously back and forth. A curious face here and there.

A shot! A cloud of smoke in front of the Third Gun. Duck! Look up again—the smoke has cleared. Reisiger waits a moment for the next shot. None comes. He pushes himself out of the hole and remains on his knees for a while. Then he stands up. Winkelmann follows him.

After a short time, everyone is gathered at the guns.

They stamp their feet to shake off the dirt. They run their greasy hands along their arms to squeeze the brown sauce out of their sleeves. You walk back and forth. At first, you don't talk. The drumbeat is still numbing your ears. But then someone laughs. And then a few laugh, briefly, embarrassed. And then, hands in dripping trouser pockets, groups stand together. "Oh, it wasn't that bad."

It wasn't that bad. At most, the water. Sitting in water for hours—it's like being in a cursed hell. Drumfire is nothing compared to that.

And people laugh again. Because jokes are made about the effort the enemy has made with no appreciable result.

The captain strolls through the position. Behind him the other officers. They are inspecting the impact.

The whole area looks as if it consists of gaping abscesses of disgustingly corroded wounds. The edges of these wounds are yellowish and have gangrenous cracks. But that's all. Here, a hair's breadth from the cover hole next to the Second Gun, three impacts are sitting on top of each other. It is inconceivable that anyone remained alive.

And the guns? The protective shield on the Third is torn open and looks like a strip of paper through which a trained dog has just jumped. A little torn. The Fourth has a deep notch at the muzzle, the length of an arm. A piece of explosive must have slipped along it. But the battery is perfectly functional. The few dead don't count.

7

There can only be one voice about it, that we would rather leave our entire eighteen army corps and forty-two million inhabitants on the battlefield than . . . In this spirit I raise my glass . . .

8

After an hour, the barrage resumes. It is concentrated on the rear area. Countless shells whistle, hiss, sing, rattle over the battery position, and hail down where 1/96 left the road yesterday and galloped onto the field. The captain's first concern: What about ammunition and provisions? How can we notify the Protzen?

Messengers cannot get through the firewall. So, there is nothing left to do but rely on the master sergeant.

The gun leaders are asked to find out what rations are still in the position. Report: a total of five bags of field rusks, three tins of meat.

Everything else has been ruined by the rain. So, at most, there was enough for the operating crew of one gun. But everyone has eaten equally little, almost nothing for twenty-four hours.

One man has the good idea to search among the dead. Maybe there's still something to be found there. They'll be soaked too, but since the dead didn't have to sit in the holes up to their necks in water, there's a chance.

Yes, next to the Second Gun lie the two flare guards and at the Fourth, the one with the belly shot, who is also dead, and each of them has two bags of rusks in their coat pockets.

The medical sergeant goes to the burnt pile of ammunition. But there's nothing there; it's all in pieces.

All in all, each man now has four to five rusks and a knifeful of beef. That lifts the spirits. Besides, there will be more at some point.

Worse than hunger now is tiredness.

One after the other lies down with their back against the hill. The ground is very wet, and the dirt from the detonation craters is a foot-high brown slurry. Nevertheless, many fall asleep.

Always above them, the buzzing and screeching arc of the shells in the air. Behind them, the roaring field.

By the willow, of which only splinters remain, the officers and Reisiger sit on ammunition baskets. Tired, tired. Stiller is asleep. Reisiger is struggling to keep his head up.

Siebert and Rossdorf discuss the situation. Something must be done! At the very least, a report to the division is necessary. In the end, 1/96 is all alone here in the middle of nowhere, and the regiment has moved off to God knows where. It's all happened before. Perhaps they also believe that none of the battery is still alive.

"If only the heavenly dog of a Wachtmeister would finally get around to sending provisions and ammunition," growls the captain. "Besides, I think it's only natural that he should show his face here."

He turns to the rear. The enemy is drumming there. He puts the binoculars to his eyes and says after a while, "I have the feeling, Rossdorf, that they're staging a real fire roll. You see ... And this wave is no doubt already rolling closer to us."

Rossdorf: "Doesn't Herr Hauptmann think we should change positions? If our battery fired the first shot, the enemy, who for sure knows where we stand, would beat us to death in five minutes."

The fire roll is getting closer.

Change of position? The hill behind, on which the battery now stands, runs to the right for another 300 meters or so. That would be the only way to change position. There is no other cover.

Yes, the fire roll is getting closer.

If you pull the guns out here and set them up 300 meters further to the right, there is at least the possibility of avoiding the target fire for the time being.

"Battery, change position! Push the guns out to the right, to where the high ground ends!"

The gunners are startled out of their sleep.

The fire roll is approaching.

The guns are pushed 300 meters further through the mud. New cover holes are dug. It's difficult; the clay weighs like a hundredweight.

Every single ammunition basket is dragged out of its drowned position. Each shell is dried and rubbed down with canvas and a coat.

The fire roll gets closer, closer.

It arrives around noon.

The 1/96 does not fire a shot. Everything is in the new holes. Hungry, tired, everyone has the feeling of being hopelessly abandoned.

The enemy drums and drums.

Around five o'clock in the afternoon, there is a direct hit in the Third Gun. A heavy caliber. Not a single screw remains intact. A pile of scrap iron.

9

German jewelry made of real projectile bronze from the battlefield. The most distinguished memento of Germany's great era.

Each piece is artistically crafted in finely chiseled handwork from the famous art workshops of the "Schule Reimann" in Berlin.

(Advertisement)

10

At around eight o'clock in the evening, infantrymen suddenly appear in the battery. Three men. Boots and uniforms dripping with mud. Their faces are black. One of them is bleeding profusely from the cheek. Blood is running down his chest. They jump over the embankment in one leap, shout "Battery Leader" and jump into the hole where Reisiger and Siebert are crouching. It is so cramped that the five men are lying on top of each other.

What's going on?

"Herr Hauptmann, we've been firing red flares for an hour. The enemy is everywhere between our lines. The battery must fire."

Siebert straightens up as best he could. "Do you have distances?"

The one with the bleeding cheek: "Our sergeant has ordered the battery to fire toward the edge of the heights. I'm sure there's no one left of us up there." He turns to the two comrades. "Well, I'm staying here. You take off, further backward. Maybe you'll get to the battalion and get reinforcements. Otherwise, that's the end of it."

The two push out of the hole and disappear.

The one with the bleeding cheek suddenly collapses and falls heavily on Reisiger.

"At least let him catch his breath," says the captain.

A shot whizzes next to the hole; half the wall collapses. Siebert ducks down. Then he straightens up—straight as a candle. "Come on, Reisiger, battery to the guns. Rapid fire one thousand."

Reisiger climbs onto the infantryman's shoulder, jumps off, and stands between the two right-hand guns. Thinks that we won't get a shot fired after all. We won't get any men out of the holes either. Then a shell hits the ground in front of him. He falls over and feels his head; everything is fine. He yells into the holes on the right and left: "Battery rapid fire one thousand!"

The two guns on the right fire. Reisiger jumps on. The second from the left is no longer visible. The pile of scrap iron has now turned to dirt. So much the better, the longer the crews will last. He shouts, "First Gun, rapid fire one thousand!"

Now the three guns are firing. It goes like clockwork. But it is uncomfortable. The correct military posture has been abandoned.

Everything is jammed close to the shields. The ammunition gunners crouch at the edge of their holes.

After a few minutes, the right wing gun stops. Reisiger, standing on the left wing, sees that the gun leader and gun layer are lying there dead. Rossdorf is now aiming instead. *There, that's all right. There's no need for a gun leader now. The captain will take care of that.*

Battery shoots and shoots and shoots at 1000.

The enemy is drumming.

It's slowly getting darker.

Yes, for God's sake, there are infantrymen in front of us again. Reisiger can clearly see that people are moving ahead. He jumps to the captain and points with his hand. "Stop firing"!

Siebert pulls the binoculars to his eyes: "They're English!"

The enemy is drumming.

The gunners of the battery stare over the shields of the three guns: Englishmen appear ten meters in front of them.

How many?

It's difficult to estimate. There are movements in various places. "Rifles!" Some are already rattling at the middle gun. The barrage is forgotten. The gunners have jumped out from behind the shield and are lying at the hill, firing.

The English raise their hands!

The enemy is drumming.

In front of the third, a direct hit hits a group that is rapidly approaching: Half a dozen men are blown up.

The others all the faster. Ten, twenty, thirty crawl over the hill. They have no more weapons. They keep raising their hands in the air as best they can against the drumfire.

Prisoners? Yes, are they prisoners?

What to do with them?

There is no time to lose; the enemy is drumming, and the battery must fire rapid fire. "Rapid fire!" yells the captain.

Then he thinks for a moment about what to do. Letting the enemy into the position might not be the right thing to do after all. *Who knows what will happen? Later, they'll stab the battery in the back?*

"Rossdorf, those guys have to get out of the position. Make sure they get away from here as quickly as possible."

Rossdorf looks at them. He feels sorry for them. They are lying directly under the barrels, exhausted, dirty, afraid. It doesn't help. He must chase them further backward, into the fire of their own batteries. Since they don't want to, since they look at him pleadingly, he kicks them. Nothing helps.

And one by one, they jump backward and disappear. Somewhere, they will collect what arrives in one piece and transport them to Zossen.

The enemy is drumming.

Siebert looks along the position. *Soon the ammunition must be out!* Order, shouted from gun to gun: "Hold your fire, everything take cover."

The captain, Rossdorf, Stiller, and Reisiger are now crouching together.

The enemy is drumming.

Where did the prisoners come from? Is this a good or a bad sign? If only we knew whether the battery was right to fire at 1000 meters. Or what's going on? And, yes, ammunition! Ammunition! Looking at the sky, it must be dark in an hour at the latest. Can we send messengers to the rear, then? Or will the master sergeant, that goddamn idiot, finally have the courtesy of his own accord? Siebert sticks his head out of the cover and pulls it back in: "Gentlemen, this all seems very bad to me. Nobody can get to us from behind. Just look at this."

Everyone looks out of the hole. The enemy is drumming.

Stiller: "Yes, Herr Hauptmann, I'm happy to give it a try."

"What?"

"To try to run up to the Protzen."

Reisiger hears this. Stiller wants to start running? Oh, Stiller wants to run off voluntarily? Well... shouldn't I... wouldn't it be my duty...?

He thinks that and dismisses it at the same moment. *No. You shouldn't do it voluntarily; I don't want to do it voluntarily anymore.*

Then he hears Rossdorf's voice: "Ammunition is coming, Herr Hauptmann."

"All heads go up! Yes, ammunition!"

Three wagons approach. One has only two horses and no men apart from the driver. One has six horses, but there is only a rider on the front horse. The rider in the middle is hanging in the stirrup, belly ripped open, head dragging in the dirt.

The gunners are already jumping out of the holes. Give us the baskets!

After a few minutes, the battery has enough shells to continue with rapid fire for a while longer.

The ammunition carts have disappeared.

Thank God there is no new order to fire. So, the battery can stay under cover. In the meantime, Siebert ponders how to determine whether a distance of 1000 meters is too much or too little.

The enemy is drumming. Only 1/96 is now spared to some extent.

Stiller has supervised the unloading of the ammunition. Reisiger stands next to him. They walk up and down.

It is almost dark. A beautiful, warm sky. A few stars. There are no clouds above the position. It's almost deep blue.

Stiller picks up a conversation where he left off days ago. "Yes, yes, Reisiger, Munich. Please, imagine that summer night on the Isar; the artists' festival; will we experience that again?"

They are at the right wing, turn around, walk toward the left, and stop where the second gun from the left was. Stiller knocks at a piece of iron. That was the gun carriage.

The two dead flare guards lie in front of them. They should at least take the identification tags off, Reisiger thinks and bends down.

Then it gets light. A crash hits his ears. He almost falls back. But then he stands up straight: "Golly, that was close, did Herr Leutnant . . ."

He notices that Stiller is swaying. He looks at him. *Is that—?*

A man is standing next to him, swaying back and forth. The man no longer has a head. A black stream shoots forward where the head used to be.

For God's sake, the body is still standing, still standing, falling forward, falling backward, yes, for God's sake . . .

The body hits the ground.

Reisiger wants to shout something: Leutnant or Herr Leutnant Stiller . . .

He rips the steel helmet off his head. Surely it can't be Stiller?

He waves the steel helmet in his hand. He realizes that his face is rainy wet with tears. He wants to jump into the next hole. He sees a face in it: It's the gunner Ziese, happily blowing the smoke of his cigarette toward him. He turns to the other hole. Oh, are they asleep? He chases off. He almost jumps on the captain's hand. He is still crying and wants to say something. Yes, Lieutenant Stiller has no head left. He just slams forward. Into the captain's lap. Says: "Lieutenant

Stiller is dead." Shakes himself in tears. He feels the captain's hand on his hair: "But Reisiger, don't be a child. I'm sorry, too. That's just the war . . . . Yes, Reisiger, the best thing will be, I've just discussed it with Lieutenant Rossdorf, we'll both go to the front. We can take another gunner with us. Let's dig a hole. So that we can finally see what's going on. It's nonsense to always shoot at 1000 meters . . . . But Reisiger, calm down. To be honest, of course, I feel sorry for little Stiller . . . . Rossdorf, you take over the battery. When we're in front, two white flares, then you'll know where we are. And then set up the telephone or relay."

"At your command, Herr Hauptmann."

Reisiger puts his steel helmet back on. I don't care about any of this. They tore Stiller's head off. So . . . they should already—

11

In close cooperation with the Volksbund[16] War Graves Commission E. V., Berlin, and the MER[17] Central European Travel Agency GmbH Berlin.

Organized individual package tours to the war graves in France and Belgium in 1926 and 1927.

A large number of inquiries from the relatives of our fallen soldiers and many hundreds of tours carried out bear eloquent witness to the great interest in such visits. The tours can now be carried out without difficulty and in such a way that the traveler finds a friendly welcome at the destination in a hotel specially selected for this purpose, that a car is available for him to visit the cemetery, and that the whole journey can be carried out at a

---

[16] Volksbund Deutsche Kriegsgräber-Fürsorge, established 1919, is responsible for the maintenance and upkeep of German war graves in Europe and North Africa. Today, the Volksbund looks after 832 German military cemeteries in 46 countries with about 2.75 million dead.

[17] MER. Mitteleuropäisches Reisebüro. Founded in 1917 by the German railway companies with rapid international expansion in the 1920s. Organized "Kraft durch Freude" (Strength Through Joy) travel in the III Reich and was involved in the Holocaust. Today, one of the leading German travel agencies under the DER brand.

moderate price, which he can pay in advance including all additional costs such as: taxes, tips, etc.

Thanks to the careful preparations, the MER package tours to the war graves do not present any difficulties for the traveler, even if the individual has little or no knowledge of the French language.

For example, it is possible to visit a cemetery about twenty kilometers away from St. Quentin from Cologne in a three-day journey—in Germany train III class, in Belgium and France II. Class—including accommodation, meals, car rides and all additional expenses at a price of eighty Reichsmark. This price is for one person; for two or more persons it would be RM Seventy-five per person. The prices give an approximate indication of the costs to be expected.

12

The enemy is drumming.

The battery has a relay to the observation post, which is about three hundred meters in front of the firing position. A hole, two meters long, fifty centimeters wide, two meters deep, was dug by Captain Siebert, Reisiger and Gunner Merkel.

Merkel was killed when the hole was dug. He lies at the edge of the pit, hunched over, a good cover against the enemy. Above his hips, the eyes of the scissor scope are glaring.

What does Siebert see? The night is torn apart, incessantly, mercilessly torn apart by a hundred thousand near and distant flames stabbing out of the earth.

An observation post? Apart from these flames, there is nothing to observe. No enemy, no friend.

The relay: A man lies in a shell crater every thirty meters. If a message comes from this observation post to the battery or from the battery to the observation post, the man jumps from one hole to the next as best he can, passes on the message, and jumps back. The second to the third and back. And the third to the fourth. A living telephone line, difficult to repair, can only be repaired with a run of twice thirty meters if the connection is interrupted somewhere by a direct hit.

The living telephone line transmits a man-to-man report to Captain Siebert: "Battery 1/96 is firing with only two guns. The left wing gun has taken a direct hit. Gunners dead."

The enemy is drumming.

Flares go up, flares all over the horizon, raspberry red flares.

Captain Siebert's order, relayed through the living telephone line to First Lieutenant Rossdorf: "Battery rapid fire. One gun at 1000 meters, one gun at 800 meters."

The message takes a long time to reach the battery.

Reisiger looks to the rear and waits and waits. *Thank God, the battery is firing!*

The enemy is drumming.

Flares go up as far as the scissor scope can be turned: raspberry red flares.

Very well. The battery fires.

The enemy is drumming.

And still, with only two guns, the battery does an excellent job. If the captain squints to the rear, the muzzle flashes blind him. It's fabulous that two guns can still achieve such firing speed.

"So, Reisiger, now look through the scissor scope. I'll rest for a moment. It's hell here, isn't it?"

Then a man of the living telephone line rushes up to the observation post: "Message from Lieutenant Rossdorf: The battery only has one gun left. The right wing gun has taken a direct hit. Gunners are dead.

First Lieutenant Rossdorf asks if he can vacate the position. He says that the Third Gun could very well stay where it is. It could be retrieved in the morning. And could Herr Hauptmann come back too?"

A second man from the living telephone line rushes up to the observation post and shouts, "Sergeant Schulz says that Lieutenant Rossdorf has just been killed. There live . . ."

The observation post brightens and rears up. The captain and Reisiger are pushed to the ground. When they get up again, the two relay posts lie dead behind them.

Now it's over.

Will there be a third report?

The enemy is drumming.

It is as bright as daylight. That's because flames are constantly leaping up right next to the observation post.

Yes, really, someone else is coming.

Reisiger grabs the captain by the sleeve.

With a terrible roar, inhuman, animal-like, a man chases through the barrage. Pushes the dead aside. A flame shoots up. Reisiger sees Winkelmann's snow-white face. He knows it exactly. It was in front of him for seven hours last night.

It is distorted.

Winkelmann rushes forward.

There is a flash.

You can see that Winkelmann is ripped open from top to bottom. His uniform is flapping to the sides, his chest is exposed, his stomach is gaping, he is carrying his intestines in his hands.

He shouts and screams.

Reisiger shouts at him: "Winkelmann . . . Karl!"

There is a flash.

Winkelmann stumbles.

He hits the ground directly above the observation hole and falls onto Reisiger.

Reisiger is doused by a warm stream.

He hears the captain say something.

The enemy is drumming.

Reisiger tries to push Winkelmann aside with his head and shoulders. Something wet and soft runs down his face.

A terrible noise!

Reisiger feels the earth rising against the sky from all sides. Then a weight lies on him. Unbearably heavy.

He wants to call out. It is impossible. He wants to move. It is impossible. He wants to open his eyes. He wants to do something.

He is freezing cold. But he can think clearly: direct hit. Buried. Life is over.

A very nice feeling. It sings and hums. All kinds of sounds in his ear. How light the earth is. The earth should be light for you. Life has come to an end. What a pity. Thank God. Thank God. The war is over. We may not be going home, but we can be sure of our Fatherland's gratitude. But that's all right now. Well, it's over—

Over? The end? Yes, that's impossible. It can't be over. I want to live. *Live! Live!*

Wide awake: *I must live.*

He realizes that he is lying with his head under the hollow of one knee. *Aha, there is air here.*

He tries to move. Impossible.

I must get my head out from between my legs first. Impossible.

He's trying to bend his knee . . . . No. He's trying to turn his foot . . . . No! If only there wasn't this hundredweight on his neck.

And what about the air? He tries to take a deep breath. But his chest is so tightly compressed.

Now he's getting hot. So, he's suffocating? So, he's buried? Buried alive and suffocating.

He turns his foot. He gives it some space. Turns his foot back and forth. Space. Space. The head becomes freer, becomes even freer. Is already between the knees. Now lift one knee to the right . . . *There, that works* . . . . Now bring your knees together. There is a hollow space. *There, that works perfectly.* He hears the earth rattling against his ears. Neck up, head firmly against the cover. Again and again. Now with his hand, push, push hard.

The hand feels air. Pull back quickly, breathe . . . *Yes, the hole has air.*

One hand through, the second behind, and now, dig and turn your hands. And bang your head. And try to stand up. Firmly press your neck against the load, then back, and give it a jerk. Then back, and a jerk. Once more, faster, faster and faster.

The earth cracks open. Reisiger's face is in the open.

The enemy is drumming.

A flame flares up in front of him.

But that is so uninteresting. He takes a deep breath. *I'm alive, steel helmet on now. And gallop, march march, move out.*

Then the lightning flashes again and hurls him to the ground. And he notices a steel helmet next to him. He grabs it and looks at it. *No, that's not mine.*

Jesus, the captain is still down below. The steel helmet belongs to the captain.

Shot. He flies a few meters to the side. Stands up half-bent over, holds his hands to his head with bent elbows. Move out as quickly as possible!

Oh no, you can't do that. The captain!

So back with the next movement. *Here is the steel helmet. Here is the crater. At one edge, the earth is thrown up; there I lay. Next to me must be the captain—*

His hands pull aside the clods to the right and left and dig deeper. His arms are up to his elbows in the hole, his fingers suddenly gripping tightly. *Thank God, it's the captain.*

Above all, a shaft has to be cleared so that the captain can breathe.

As the next shot is fired, close to him, and the darkness is torn apart, he sees Siebert's face. Yes, it is free. If he is still alive at all, he can breathe.

The hands continue to dig, exposing the head, the hair, and finally reaching along the back under the armpits.

"Herr Hauptmann!" He tugs him and tries to pull the body out. It is difficult. Every time he stands up, he has to drop the load because fire and explosives are flying over him.

Finally, with a jerk, out comes the upper body! Reisiger lies with his arms stretched out in front of him, his hands folded over the captain's chest. He crawls backward on his stomach, dragging his body behind him. This is very difficult. A lot of time passes before he has him on level ground.

Reisiger no longer pays attention to the shots. His fear takes a different path. The feeling of lying here against the dawning morning, completely abandoned as the only living person in a field, gives him a fright: The captain is dead.

He shouts, he pleads, "Herr Hauptmann." And again and again, imploringly, "Herr Hauptmann."

He doesn't move. Lying like a wet sack, heavy, limp.

Reisiger tries to grab his hands and tries to lift an arm. Everything claps to the ground again and again.

The enemy is drumming.

It's a gray morning sky. How much better the darkness was. Now you can see all the horror much more clearly, overly clear and distorted by the pale light and supernaturally magnified within all limits.

Reisiger ponders: *How good it must be to run toward the next shot with arms outstretched, to hold your whole body out, to simply let yourself be torn to shreds.*

He thinks about it: *The whole battery is dead. It's only decent if I lie down next to my dead captain and wait for it to finally hit me too.*

But when a shot crashes down, sending poor Winkelmann's torn body spinning into the air, he is gripped by fear again. And the wish, the call, the command that suddenly seems to him the highest command: To live!

He rushes furiously at Siebert, kneels down close to him, and strikes him furiously on the chest with both fists. His fists clatter senselessly, and he screams louder and louder, ever more shrilly, "Hauptmann! Hauptmann Siebert! Herr Hauptmann! Hauptmann!"

Then he sits astride the body, grabs Siebert's arms firmly by the wrists, bends them, then stretches them. He knows that this is the way to bring the drowned to life. He has to force it here too.

Siebert's face is swollen, a blue-red. His mouth is encrusted with earth. *The eyes—are the eyes closed?* Hasn't one eyelid just fluttered open and closed? Reisiger bends down. *Yes, there must be a trembling in the eyelids.*

He rips open Siebert's skirt, tears open his shirt, and feels with both hands. *Yes, there is warmth in his body.* Presses his ear against his chest. A world of its own, not yet suffocated, not to be drowned out by any drumfire, the heart beats . . . Jerky, very irregular, violent.

The captain is alive.

Reisiger screams louder, again and again.

He notices the sweat burning down his face.

A shot, very close, and another shot: They've woken Siebert! He suddenly opens his eyes and straightens up at the same moment. Another shot. How strangely a human being works. The newly awakened captain ducks and takes cover according to the rules, habitually. And as the splinters whiz past, he rises carefully and then says: "Reisiger, I think it's time we changed positions." He smiles, a smile, very embarrassed, a little tired.

"Does it hurt Herr Hauptmann anywhere?"

"But never mind . . . you see, Reisiger: we're still alive." Again, the embarrassed smile, which then changes abruptly. The face is distorted, desperate. "Reisiger, where's our battery? Come on, come on, we must get into the firing position!"

Machine guns are now clattering into the enemy's artillery fire. It's not far from the observation hole to the firing position. But the jump from funnel to funnel is slowing us down. Not only do they have to wait for the buzzing of the shrapnel, but they also have to be careful not to be sawn to pieces by bullets.

And weren't there a few hundredweight of clay on Siebert and Reisiger just a short time ago? All their bones ached; their chests were constricted as if their ribs had been pressed into their lungs. Breathing is difficult. All it takes is a run of ten steps and a red haze falls over their eyes. Siebert jumps forward and is already in the next funnel; Reisiger wants to follow, staggers, sinks back, loses his senses, and only wakes up when the captain turns back and calls him.

This goes back and forth several times.

The day is bright now, morning air. Both have their skirts unbuttoned, their chests free for a little cooling.

One jump, lie down, one jump, be tossed around. And many jumps. And not a single shot is heard against the enemy.

And between jump and jump, the certainty of being alone, the enemy in front, nothing behind.

And then they both stand where the willow once was. Crawling through the former firing position. Thank God they are covered a little by the hill.

But even crawling is torture. Siebert straightens up and starts running so that Reisiger can hardly follow him.

And then: 1/96.

There is no way to rest for even a few seconds here. Here, it is worse than lying alone between craters and gunfire in an open field.

They arrive on the left wing. They rush through the position, hardly daring to look up, only hopping carefully so as not to step on the dead.

Siebert in front, Reisiger behind.

Familiar faces everywhere. Fat Herbst, God, he was funny, especially when he was drunk. Then he grinned all the way to his ears. Now, his face is strangely white, and his eyes look desperately at the sky. There is, once again, Lieutenant Stiller. Reisiger walks past him with his eyes closed and his teeth clenched. It must be him; you can see the epaulets. *And on and on.* Now already at the Third Gun, and no one is alive. If only someone were wounded, if only someone raised a hand or moved a leg. If only someone could hear a voice.

Past Lieutenant Rossdorf, just in front of the Fourth Gun. His forehead is deeply furrowed, and he still has a flare gun in his hand.

And at the Fourth Gun, Sergeant Schulz. The shoulders are bent, very correct and dashing, his right hand on the trigger; the gun could fire at any moment. But his skirt is torn open on the left. The blood is still dripping.

This gun will not fire another shot in this war. The captain has already moved on. And Reisiger after him.

The stretch of cover comes to an end. Now comes the endless deserted open field again. Onward, onward, onward.

Where to? The captain doesn't seem to know himself. It is difficult to determine the direction. He stands for a moment, jumps a few more steps, throws himself down.

Behind him, separated by the column of smoke from a heavy shot, lies Reisiger.

Thank God, he thinks, *now the captain is dead!*

Now I have no obligation to run a single step further.

His chest is strangely hot, as if something burning is rising inside him. As he runs the back of his hand over his mouth and nose, he sees that he is bleeding from his mouth and nose.

Also good. It doesn't matter how you die here. The lungs must be on strike. It must be the result of having been buried.

The smoke has disappeared. The captain stands up, waves, walks on.

Reisiger, blood from his mouth and nose, runs on.

An area spared from artillery fire. Maybe 100 meters, unthinkable, without smoke. A veritable meadow, deeply rutted but all the more beautiful, between the blast funnels white clover, ah, and meadowfoam. Nice that it's summer.

A buzzing in the air, thunder, we know that. Reisiger jumps on, crashes into the captain, and both fall into a funnel. "Easy, Reisiger, don't move, another aviator."

But they put their heads up a little. The plane above them is almost close enough to reach with your hands. Ridiculously low. They see a broad laughing face leaning out from behind the pilot's seat, and see two hands tugging at the machine gun. And as the monster pushes past, a few bullets clap at their feet.

"Bloody bastards," says Siebert, "they seem to have it in abundance if they can afford to shoot at every single person with airplanes. If only the dog doesn't come back."

They storm on.

This time Reisiger leads the way. Where to? You can only tell by the impact of the bullets that you're not running into the enemy.

No, it's definitely going backward. The barrage is letting up a bit back here. Maybe reserves will finally arrive; maybe you'll come across German soldiers somewhere.

Reisiger hears his name being called. Siebert waves to him and walks off at a right angle.

There is a mound slightly raised out of the ground. After a few steps, you recognize it clearly. A concrete hilltop; it must be a bunker. Is there infantry there?

A shot goes off over their heads. They fall to the ground and see that the beast hits the concrete block and bounces off because it can't crush the cover.

After a few jumps, they stand in front of a staircase that leads deep into the ground. Siebert calls out. A voice answers from below: "Hello, who's there? This is the headquarters of the 2nd Battalion Infantry 18."

More blood pours out of Reisiger's mouth and nose. His skirt and trousers are completely soaked. His face and hands are red.

One shot! They both fly down a dozen steps, roll over each other, and lie on the floor, distraught and astonished. In front of them sits an infantry major, cigar in his mouth, and next to him, two younger officers.

Siebert stands up, wants to put his hand on his helmet, realizes that he doesn't have one on, mumbles something, and announces: "Captain Siebert, Battery Leader 1st F.A.R. 96. Herr Major, we two are the only survivors of the battery."

He points to Reisiger. Reisiger also wants to stand up. But if he moves just a little, blood shoots out of his nose like a bath, and he has to stay down.

The major: "Well, what's wrong?"

Siebert: "We were buried, Herr Major."

Reisiger only hears this very faintly. It seems to him that he is sealed off from the outside world by a thick curtain. The curtain closes tightly. For a while, he can only feel that he is being pushed into the dark. Then he opens his eyes because he hears his name. The captain kneels beside him. "Reisiger, you stay here. I have spoken to the major. You will be taken away as soon as possible. I'm looking for the regiment. And goodbye. The war will be over for you."

Reisiger straightens up. *War over? No, no, it's not that simple. Of course, I'm going with him . . . .* He straightens up, sits, kneels, stands;

excellent! The blood from his nose is not so unpleasant. If only it didn't keep rising in his throat and filling his mouth. He says with difficulty, "I'll go with you, Herr Hauptmann."

The major waves him off, very gruffly.

The captain shakes his head. He shakes hands with the major, the two other officers, Reisiger. Then he jumps up the stairs.

Reisiger wants to follow. Impossible. He stops at the doorway. He looks after the captain, still the knees, the gaiters, still the boot heels—gone.

Then, thank God, his voice again: "Reisiger, give me your steel helmet. I left mine behind. You won't be needing it after all. Wait, I'm coming."

Reisiger rips the helmet off his head. *Oh no, I'll manage the stairs.* He goes up step by step.

The sun is shining beautifully.

He hands Siebert the steel helmet. He grabs it.

A shot cracks, and the whole bunker rises. A scream. The steel helmet flies past Reisiger into the depths. The captain's torn right hand hangs tightly from the leather strap.

The captain rushes past Reisiger, down the stairs after the helmet and his hand.

Then the black curtain is there again. How violently it comes down. Over the sun and over the blue sky.

Thoughts whirl through Reisiger: *The torn-off hand of my captain gives me back my steel helmet. Haha, a salute! Helmet off, hand off for prayer ... Can't give you my hand ... good comrade ... a splendid cantus ex est ...*

Then all is dark.

13

According to the latest available statistics, 90.2 percent of the members of the German field army treated in the military hospitals throughout the German homeland were fit for service again, 1.4 percent died, and 8.4 percent remained unfit for service or were granted leave. (WTB)

14

Grand Headquarters. War reports May/June:

In the last two months, the general war situation has steadily worsened to such an extent that the turn from June to July seems less suitable than ever for a summarizing review.

It is not the first time that a complete reversal of the situation to our disadvantage has been announced by the entire press of our opponents long before the events that were to bring it about.

Neither these announcements nor the deeds that followed them have ever been able to dispel our calm.

15

The age of officers may be stated in obituaries without objection, provided that No. 23 O.Z. is observed. Inadmissible: eighteen-year-old lieutenant as company commander.

(Chief Censor's Office No. 375. O.Z. May, 15)

Obituaries:

1. Long collective obituaries of troop units are prohibited; maximum five to six names.
2. Collective obituaries of associations & clubs etc. are permitted. (No. 23 O.Z. December 4, 14)

Chapter III

1

The War Food Office writes:

With the shortage of fat, soap, and lights, a voluntary restriction in the use of Christmas candles is urgently required this year. It would be best if only one candle were lit on each Christmas tree. This would in no way detract from the significance and solemnity of the occasion. For the children, however, for whom the Christmas trees are mainly intended, it will remain a valuable memory for their whole lives that only one candle was allowed to burn on their tree in the war year 1916.

(*Vossische Zeitung*, Berlin, December 7, 1916)

2

The directors of the German Potato Cultivation Station, which is affiliated with the Institute for the Fermentation Trade, had a stimulating exchange of messages these days. This was because the "Hindenburg"[18] variety had come out on top in the examination of this year's results of the German Potato Cultivation Station. The following telegram was sent to Field Marshal Hindenburg: "We respectfully report to Your Excellency that the newly bred potato series of Mr. Kamecke-Streckenthin 'Hindenburg' has won against nineteen other varieties in the average of the thirty test fields distributed throughout the German Reich with the brilliant yield of 279.1

---

[18] Paul von Hindenburg, 1847–1934. Prussian Field Marshal. Became a national hero after the victorious battle of Tannenberg in 1914. Chief of the Great General Staff 1916–1918. President of the Weimar Republic 1925–1934. Hindenburg agreed in January 1933 to appoint Hitler as chancellor.

quintals and 50 quintals of starch per hectare. The home front of German agriculture did not rest; the potato harvest, which increased again under the leadership of 'Hindenburg', ensured food for the people and the army. For the German Potato Cultivation Station of the Association of Spirit Producers in Germany at the Institute for Fermentation Purposes. A. Säuberlich, M. Delbrück, C. v. Eckenbrecher"

General Field Marshal v. Hindenburg sent the following reply:

"Thank you very much for the kind news about the newly bred 'Hindenburg' potato variety, which I was pleased to receive. I know what we owe to the successful homework of German agriculture."

(*Hamburger Tageblatt*, February 19, 1917)

3

"... I was pleased ..."

4

For some time now I have been able to observe the splendid process of distancing miles and miles of our former front line from the enemy here on the Aisne front in all its preparations, and in the last few days I have witnessed the completely loss-free withdrawal of enormous bodies of troops from the front line to every new receiving position ...

The destruction of all the positions left behind, the roads, bridges, railroads, wire lines, the laying down of all cover and shelters has been carried out with great thoroughness by the demolition detachments ...

There is no need to lament the picturesque monuments of the enemy's country that have been destroyed in the process, now that the enemy has once again expressed the will to destroy Germany. No perishable good in the world has a right to exist if it stands in the way of German victory.

(Scheuermann, An der Aisne, 19. 3. *Deutsche Tageszeitung*, Berlin, March 20, 1917)

5

Every opportunity must be taken to prove that German warfare avoids all unnecessary hardships. In enemy towns and villages, the most important monuments are to be photographed in such a way that their integrity can be proven. Always photograph a few German soldiers as well. Enemy devastation and atrocities are to be proven by pictures.

(General Staff, 28. 12. 1914. III. C. 3094. Pr. excerpt from the instructions for war photographers and cinematographers)

6

Ten Proverbs on the War Loan

1. Everyone in the German nation must give,
So that the clouds of war may not outlive.

2. If you want to help win the bloody struggle,
You must bring your fighters your coins double.

3. Our good boys out there in the fields
are worth more than all your goods and yields.

4. Give freely and willingly to the Fatherland now,
Or else you must give to the enemy despite your vow.

5. If the enemy is in the land where you were born,
Then thy penny is lost and all is torn.

6. Do not let yourself be beguiled by a grumbling mind;
You make no sacrifice at all; you create gain for yourself.

7. He who guards his wealth at the darkest hour,
He drags himself to doom with a starving power.

8. There is no choice here; there is no fray,
With pleasure and love we sign and pay!

9. Sing: "Heart to heart and hand in hand!"
Now show your love for the Fatherland!

10. *Let your hearts glow and open your hand!*
You will all help to the victorious end!

(H. Weinert, *Neueste Nachrichten*,
Braunschweig, March 31, 1917)

7

Reisiger was in a military hospital in the Harz Mountains until mid-November 1916. He was then transferred to the replacement section of his regiment and was written fit for field service at the beginning of December. Promoted to deputy officer on December 23.

8

Lieutenant Römer, the battery commander of the Ninth Battery of the Reserve Field Artillery Regiment 253, receives the order to submit a report to the regimental staff on the suitability of the deputy officer Reisiger 9/253 as a reserve officer.

He writes:

Deputy Officer Reisiger was transferred to the Ninth Battery from the Reserve Battalion on December 31, 1916. R. has been serving as an officer from that time until today. He was detached to the Rembertow firing school near Warsaw during the months of June and July 1917. The regiment has received good reports.

I consider R. to be a capable, energetic superior who knows how to maintain discipline among his subordinates.

R.'s shooting knowledge is good. He alternates with Lieutenant Sauer on observation duty and has had several opportunities to shoot independently with the battery.

The ballots of all officers of the division showed a plus for him.

I questioned R. personally about his political views. (I am referring here to the meeting that the regimental commander had with me concerning R.'s alleged collaboration on a Social Democratic newspaper or magazine). R. expressly emphasizes that he is of the opinion that this collaboration can only be understood as artistic. He gave me two poems published in a Berlin weekly, which are enclosed.

Despite everything, I would like to warmly endorse his promotion to lieutenant in the reserve.

Römer, Ltn. and Battery Commander.

Attachments: Two poems by the deputy officer Reisiger, published at the end of 1916 in the magazine "Aktion."

Dreaming

Dreaming
My rifle is petrified—like yours, my brother,
We no longer aim at each other.
Flowers grow on the stock and barrel;
red flowers.
(Oh! that they still remind us of murder!)
The trenches that ate us in fear
are rising. And become soft and level.
We see light—land—us.
We feel: Human.
And kiss each other.
Oh!
Dream!

Loretto (For H. K.)

To drown in silence for a day!
Cool your head in flowers for a day
and let your hands fall
and dream: this black velvet, singing dream:
For one day, not to kill.

9

Firing position 9/253 has been located for many months, not far from Postawy, northwest of Vilnius. The guns are in a grove behind thick palisades made of tree trunks, and the crews and officers in spacious log cabins. Reisiger lives in a small hut with Master Sergeant Spilcker. He has to eat his meals in the battery commander's house. Lieutenant Sauer, the only battery officer, also eats with them.

In the evening, when Römer has sent his report to the regiment, instead of the usual game of cards, which Reisiger has to take part in every day from nine o'clock until about twelve o'clock at night, when he is not on observation, Lieutenant Sauer is asked to leave the room for a while; Römer has to talk to Reisiger.

Reisiger sits at the table; Römer walks back and forth in front of him, puffs from a pipe, and doesn't talk. Finally, he stands against the stove and takes from his coat pocket the two poems that Reisiger was forced to hand him a few days ago.

He begins a conversation.

"Reisiger, you know that I mean well by you; you will not be able to complain that I have ever treated you badly as long as you are here in the battery. On the contrary, I have always taken pains never to forget that we are, in a sense, fellow students in peacetime . . ." The lieutenant pauses and paces back and forth.

Reisiger doesn't quite know what the conversation is about. He finally recognizes the newspaper clippings Römer is holding: "Herr Leutnant, do you want to talk about my poems?"

"Yes." The lieutenant pulls up a chair, places the poems on the table, and leans on his elbows: "If that goes well."

Reisiger knows nothing to say. Pause. After a while, the lieutenant taps one of the two pieces of paper with his finger and says, "No killing for a day? Are you a pacifist?"

Reisiger, very honest: "I haven't thought about that yet, Herr Leutnant." *Pacifist*, he thinks. *I don't even think I'm a pacifist.* "I just think . . ."

The lieutenant interrupts him: "So even if you want to tell me that your poems have nothing to do with politics . . . I doubt that anyone higher up will believe that . . . . Reisiger, let's not kid ourselves, you can talk to me, honestly . . . what's going on? What were you thinking?

You're not a child anymore . . . it must make sense if you write a sentence like 'Don't kill for a day' or . . ." He taps his pipe on the other page. "Here: 'We no longer aim at each other.' What are you thinking?"

The moment Reisiger speaks, he realizes that it is wrong to speak. He says anyway, "Herr Leutnant . . . maybe . . . maybe . . . I can't . . . or I don't like it anymore."

So, now it's out.

How quiet it is in the room. The lieutenant sucks on his pipe. It makes a whistling sound every time. Sounds very funny. But he also puffs harder than ever. The clouds of smoke are always more voluminous. It stinks.

Then he gets up, stands with his legs apart, his arms straddling the tabletop, stares into the air, and says, "Hmm . . . Reisiger, I think the good Lord has abandoned you. You are, it seems to me, already completely mad. But golly, how can you imagine that you'll ever become an officer if you say: 'I can't or I don't like it anymore.' That's nonsense, senseless nonsense! And even if you're serious . . . Reisiger, you can't say that. Honestly, has anyone ever asked you, I mean you privately, to kill? I mean, we artillerymen have it particularly good in this respect. We usually shoot without even seeing where it hits and who it hits. Right? Well, then you don't really care. And by the way, if you don't kill yourself, you *will* be killed. Well, would you prefer that?"

There's the question again. Reisiger has thought about this a thousand times. If you don't kill, you will be killed.

Would I prefer that? No.

That's the most horrible question in war: Would you rather prefer, when an enemy is standing in front of you and his bayonet is in front of you, that he runs it into your stomach or would you rather hit him on the skull? And live?

"Excuse me, Herr Leutnant . . . I don't know."

He would like to add . . . *Maybe it's better if I don't become an officer. Because I think it's terrible to the point of madness to have to answer for whether the hundred people who blindly obey my orders kill or are killed. And that's what it comes down to.*

Then the lieutenant takes the two newspaper clippings, slowly tears them in half, and says with a small smile, "Reisiger, you may be crazy, but now do me a favor and don't plunge yourself into disaster. Leave that nonsense alone. What do you have to write? Do you have to suffer here? You must admit to me that God knows I'm not speaking to you in an official capacity. Because I know very well that you are a decent soldier. Don't make trouble for yourself, and don't make trouble for me. I can tell you in private that all the gentlemen in the unit have approved your election as lieutenant. So why all this fuss? It may well be that the war will soon be over. There you go. At home, you can write as much poetry as you like . . . And now I'm talking to you on official business . . . I want the nonsense to stop. Do you understand me?"

Reisiger stands up, stands at attention. "At your command, Herr Leutnant."

"Don't embarrass me. You've been in the field since '14 too. It's not decent for an old soldier to suddenly go limp after such a long time."

Römer's voice becomes hard and angry. "By the way, don't forget that a stupid excuse like the one you used just now might be enough to get you put up against the wall or sent to prison."

He holds out his hand to Reisiger. "So you promise me you'll stop writing these childish poems."

Reisiger, after a while, puts his hand in the lieutenant's. "By your command, Herr Leutnant."

Römer: "Well, I think, Reisiger, that your promotion will come in two weeks. And then it will be easier for you anyway."

Reisiger, looking into Römer's eyes, says, "Thank you obediently, Herr Leutnant."

The lieutenant laughs: "Lord God, how solemn. I feel downright funny that I have to talk you into becoming an officer like a sick horse . . . . By the way, you know that the Iron Cross ribbon is worn in the second buttonhole. You have it in the third. It looks like a club badge. Please, have the goodness to change that too . . . . So, we understand each other?"

He calls into the kitchen to the lad: "Urban, bring three bottles of beer and call Lieutenant Sauer."

Then he turns back to Reisiger: "There's nothing more stupid than poets . . . Go, go, Reisiger, shuffle, shuffle. The skat has to go up, otherwise we'll all miss out tonight."

Reisiger shuffles the cards. The lad Urban brings three bottles of beer. Lieutenant Sauer enters, takes off his skirt, and sits in pants with black-and-white suspenders over a sticky Lahmann shirt. They play skat, with pros and cons. Reisiger wins until around half-past two o'clock in the morning, seven marks thirty pfennigs. Then he goes to sleep.

10

Turn your money into submarines, barbed wire, guns and grenades, machine guns and bullets, and you will save the lives of our heroes at the front. It is necessary to prove to our enemies by the result of the war loan that Germany's economic strength is undiminished, so that they lose the courage and hope of ever being able to defeat us.
(Warsaw newspaper, March 15, 1917)

11

To the Heads of State of the Belligerent Nations:

Towards the end of the first year of the war, We addressed the nations at war with the most lively exhortations and also indicated the path to be followed in order to achieve a lasting and honorable peace for all. Unfortunately, Our call was not heard, and the war continued bitterly for two years with all its horrors; it became even more cruel and spread by land and sea, even into the air; devastation and death were seen to befall undefended cities, quiet villages and their innocent populations. And now, no one can imagine how much the suffering of all would be increased and aggravated if further months, or worse still, further years, were to follow these bloody three years. Is the civilized world to become a field of death? Is Europe, so glorious and prosperous, as if carried away by a general madness, to rush towards the abyss and lend a hand to its own destruction?

Vatican, August 1, 1917

Benedictus PPXV

12

We fence and strike until the opponent has had enough . . .
(Wilhelm II, after G. Queri, *Berliner Tageblatt*, August 23, 1917)

No people, therefore, have more reason than the Germans to wish that a conciliatory and fraternal spirit between the nations will take the place of general hatred and fighting. (Reply to the Pope, Berlin, September 19, 1917)

. . . turn your money into submarines . . .

13

When strong men, as the Central Powers, thank God, are, are attacked by a band of robbers who want to strangle them, then they defend themselves and, if they can, beat their opponents down completely—but they do not negotiate with them, that would also be completely useless. But that is our case, and therefore, no agreement, no negotiation, no peace conference, but only complete victory, complete defeat of all opponents, and then dictate to everyone the peace that we can and want to grant.

(Lieutenant General z. D. v. Roon, Alldeutsche Korrespondenz, according to *Berliner Tageblatt*, August 25, 1917)

14

I regard the peace resolution of the so-called Reichstag majority as a betrayal of the Fatherland, and if it were to come to a referendum, the result would probably be quite different.

(Letter from a worker, *Deutsche Tageszeitung*, August 9, 1917)

15

. . . We must await some new evidence of the intentions of the great peoples of the Central Powers. God grant that these will be given soon and in such a way that they will re-establish the trust of all peoples in the good faith of the nations and the possibility of a peace concluded by treaty.

Robert Lansing,
Secretary of State

16

Let us go through this hour with the old character of our race, and next spring, we will begin, the world will begin, to reap the fruits of our valor.

(Lloyd George, House of Commons, August 16, 1917)

17

It is, of course, impossible to prevent even perfectly reasonable and unobjectionable articles in the German press from being distorted and distorted by the enemy. But we must always bear in mind that our opponents act according to this recipe. It is worth noting that all articles abroad that are based on a longing for peace are extremely damaging to us. Our opponents are so mentally misled by incitement and lies that, despite all their failures, they still believe in their victory and interpret everything that looks like hope for peace as war weariness, weakness, and exhaustion. Peace articles only weaken the moral effect of our military successes.

(Censorship book for the German press, March 1917, cf. peace issue)

18

The papal rally has once again placed the nations of Europe at a crossroads. Once again, before the decisive winter campaign, they have been given the opportunity to begin the reconstruction of Europe, bleeding from deep wounds but with a bare shield. It is now up to Germany's opponents to prove whether they too have caught a whiff of the new spirit . . . . Thus, in this decisive hour of destiny, the German people stand strong but silent, powerful but moderate, ready to fight as only ever before, but also ready to help realize the word of peace on earth.

(v. Kühlmann, Reichstag, October 23, 1917)

19

The government of workers and peasants created by the revolution of November 6 and 7, which is based on the Workers' and Soldiers' Council, proposes to all governments of the belligerents to begin discussions on a just and democratic peace as soon as possible. . . The government proposes to the governments of all the belligerent countries that an armistice be concluded immediately. For its part, it believes that this armistice must be concluded for three months . . . (The Soviet to all belligerent States)

Berlin, November 23 (unofficial):

The order of the Maximalist rulers in Petersburg to initiate an armistice has been rejected by the commander-in-chief of the Russian army, according to a radio message intercepted at the front.

To the peoples of the belligerent countries!

We, the Council of People's Commissars, address this question to the governments of our allies: France, Great Britain, Italy, the United States, Belgium, Serbia, Romania, Japan and China. We ask them in the face of their own people, in the face of the whole world, whether they agree to enter into peace negotiations . . . We ask the people whether reactionary diplomacy is expressing its thoughts and aspirations, whether the people will allow diplomacy to abandon the great opportunity for peace opened up by the Russian revolution. The answer to this question . . (disorder) . . . down with the winter campaign! Long live peace and the brotherhood of nations!

*The People's Commissar for Foreign Affairs,
Trotsky*
(Garbled radio message, Tsarskoye Sselo, November. 28)

20

Gentlemen! At this time our eyes are directed above all to the East. Russia, which has hurled the torch of war into the world, Russia, in which a rotten-to-the-bone pack of bureaucrats and parasites pushed aside a perhaps sometimes well-meaning, but weak and misguided autocrat, had deceived the mobilization, which has become the actual and direct cause of this enormous nations' catastrophe, has swept away the culprits and is now struggling hard to gain room for its reconstruction through truce and peace.

(v. Kühlmann, Reichstag, November 30, 1917)

Chapter IV

1

December 3, 1917　　　　　　　　　　　*Grand Headquarters*

Eastern Theater of War

In numerous sections of the Russian front, local truces have been agreed from division to division.

2

Letter from Lieutenant Reisiger to his parents:

Artillery Firing position 9/253, December 6, 1917

Dear parents!

By the time you receive this letter, you will have heard that the truce has now "broken out" here. We have just received a telephone message that we are obliged not to fire anymore. That was to be expected for days, but nobody wanted to believe it until the last moment.

　　It's now three o'clock in the afternoon; it's insanely cold. I'm sitting with my sergeant and a telephone operator on the observation post close behind our first trench. When I look through the scissor scope, I realize what truce means. The German and Russian trenches are about 200 meters apart. The soldiers have crawled out of all the positions. Wherever you look, the German infantrymen are standing together with the Russian soldiers in an open field. Nobody has any weapons with them. I heard earlier that a lively barter trade has started instead. Our infantrymen are giving away cigarettes for soap and Russian tea.

I'm almost envious when I see it. I would love to sneak out of the observation here and try to talk to a Russian myself. The feeling that the "enemies" have now suddenly become human beings, because some of the high gentlemen wanted it that way, is wonderful.

I have to stay up here tonight, but I don't know why. Tomorrow morning, however, I'm sure I'll be strolling around between the trenches with the infantrymen. It's an unimaginable thought that you can really walk on two legs, upright and high on the bare earth, without the feeling that the enemy is shooting you between the eyes.

Farewell.

I became an officer a quarter of a year ago yesterday. An important day; will Mother think of it? (It was nicer, by the way, being a real "common" soldier. Less because of playing skat but more because you have to be so terribly fine. I shi . . . on the finesse). And what is France doing? Dear God! I'm bloody sick of it, but an officer isn't allowed to say that.

The postal embargo that some kind gentleman had imposed on me (presumably because of my war poems) seems to have been graciously lifted. Yesterday I received a letter from Mother that had not been "opened by the authorities." So, you have a good son after all.

Your Adolf

3

Fixed "visiting times" are established between Russians and Germans when the soldiers meet each other in front of the trenches. How quickly one gets used to no longer seeing an enemy in the "enemy!" Bartering flourished, but even without goods, the men stood together, spoke friendly words, and tried to forget, to negate the war. And at night, when everyone is back in their holes, songs are sung which, despite all the sentimentality, express something like happiness about the buried hostilities.

Until shortly before Christmas, a divisional order reaches the German positions: All future conversations and encounters with the Russians are strictly forbidden.

Reisiger learns of this at the observation post, which still has to be manned by an officer, a sergeant and a telephonist on a daily basis.

He is sitting at the scissor scope; it is almost midday. There is a faint mist over the area, which the sun cannot penetrate. The temperature is minus twenty-seven degrees.

The visibility is poor. The scissor scope has been set up for days in direction of a group of trees roughly in the middle between the positions. Here, despite the hazy air, the eyes find what they are looking for: the Russians, who arrive at the usual time. Four bearded men arrive first. In green overalls, without headgear, with long shaggy hair. They look around, talk to each other, and finally crouch down between the tree trunks despite the cold. Waiting.

After a while more come; tall, young people with wide leather belts, soft boots that almost reach their stomachs, gray caps or high helmets made of fur. They, too, look toward the German trench. Waiting.

Waiting as one can only wait in Russia. They step from one leg to the other, squat down, stand up, and speak at first. Then they fall silent.

The scissor scope shows their faces almost in natural size.

Reisiger sees how the bearded and smooth ones become more and more immobile.

Then many minutes pass in which no more mouths open. The faces no longer look at each other either, they stare at the floor.

Twelve noon used to be the usual meeting time, but now it's almost three.

Reisiger forgets that he is sitting in the observation post of his battery at the scissor scope. He feels part of the group there. It's all so close at hand. It almost tempts him to say a word or to explain, to apologize, to make himself understood at all. To make it clear to the Russians that it wasn't the soldiers who were guilty. The order—

It starts to snow. The bearded men were still sitting on the ground, hands folded over their knees, staring.

Finally, they move again. The bearded men stand up and speak to their younger comrades. One points with his hand; another puts both hands to his mouth and shouts against the German position. No one answers. Another conversation begins among the group. Then two of the younger men move forward, hesitantly, toward the German trench.

Reisiger's heart is pounding. What will happen? Any approach, any contact with the Russians is strictly forbidden.

What can happen? In the German trench, as before, the posts stand, as before, the rifle lies in the embrasure. Is it loaded?

The pounding in his heart grows. A childish idea rises in Reisiger's mind: The truce could now be torn apart with an insane rifle shot.

The two Russians come closer and closer. Reisiger follows them. Another ten steps, eight steps, five steps in front of the German position.

The scissor scope shifts back to the large group; there they stand, the bearded men and the boys, their eyes fixed on the two comrades.

And back to them, five paces in front of the German position; nothing moves. The driving snow gets heavier, the twilight darker. You can see one of them take off his high fur hat and turn it in his hands. You can see him calling out. Everything remains quiet, motionless.

The two Russians slowly turn around and walk back to the group.

The group moves away. Two and two in order, strangely solemn, as if behind a funeral. The group disappears.

Reisiger's heart races through his throat. He turns the scope to the German position. And? The guards are standing there in the driving snow, their elbows resting on the breastworks, their eyes boring into the retreating Russians.

Reisiger's head is filled with rage. Can't you feel the desolation in the infantrymen's posture? How crouched they are! How paralyzed they are! How they look after the Russians: as if they were losing hope. But they must not hold them back. Divisional order. Armistice with divisional orders: Talking to the enemy is forbidden.

He turns the scissor scope with the glasses into the observation post. He looks through it again; everything is black.

And back: no more Russians to be seen.

4

We receive reports from Bayreuth: The coal shortage is also having an effect at Villa Wahnfried. Mrs. Cosima Wagner, who incidentally enjoys an astonishing level of sprightliness and who can be seen taking her daily walks even in the worst weather, celebrates her 80th birthday on Boxing Day. In the last issue of the "Oberfränkische Zeitung," Siegfried Wagner published the following request: "As the reception rooms in Wahnfried cannot be heated due to a lack of coal, we must, to our deep regret, ask our friends to kindly refrain from personally congratulating our mother on her 80th birthday. Siegfried Wagner and family."

(*Vossische Zeitung*, Berlin, 21. 12. 1917)

5

The armistice quickly became part of everyday life.

When you walk through the neighborhoods in the evening, the word "armistice" comes up from time to time. But it seems as if its glow is no longer alive. It's uttered through the teeth, angrily, mockingly. You wave your hand away: "Just as garbage as the war .... But they can do it with us."

On Christmas Eve, Lieutenant Römer gathers the battery together as darkness falls and gives a speech: "It doesn't help, we have to endure. I admit that this is slowly leading to madness. You know, boys, I'm no orator, but if we're already celebrating Christmas here in the field, then let me say one thing: The decent soldier does his duty wherever he's put. We'll get through the truce and then . . . peace on earth."

People drink Glühwein. There are five cigars, ten cigarettes and chocolate per head from a Red Cross gift box. There is one large piece of pork per gun. Candles burn on small fir trees in all the blockhouses.

Römer, Sauer, Reisiger, and Sergeant Spilcker are sitting in the officers' quarters. Roast pork, Glühwein, a bottle of sparkling wine. Then they play skat. A festive game of skat that lasts until around four in the morning.

Early on Christmas Day, the regiment receives its orders by telephone: Lieutenant Reisiger is transferred to the staff of the III Battalion as an ordnance officer. Report at one o'clock noon.

6

The leader of the III Battalion was the most feared officer in the regiment. Captain Brett, reserve officer. So rude that even the regimental commander only spoke to him personally when necessary. But probably the best officer in the regiment. The most reliable expert in all technical shooting matters, tactically clever, organizationally a role model.

Reisiger couldn't understand why he had been chosen as ordnance officer. He was not feeling very well when he reported to staff quarters at one o'clock in the afternoon on Christmas Day. The gentlemen from the staff were at lunch. Very nice. When Reisiger entered the room, there was a roast goose on the table. Captain Brett listened to his report, then even stood up, a huge napkin around his neck, shook his hand, and introduced himself: "This is the adjutant, Lieutenant Weller, this is the battalion doctor, Field Surgeon Winkel. Have you eaten yet? So take a seat."

Then everyone fell silent.

The captain ate endless portions, resting one hand on the table and occasionally leafing through a book next to him with the other.

The adjutant and doctor, who were sitting opposite each other, grinned silently at each other from time to time. Strange company.

When the captain had stuffed the last morsel into his mouth, he stood up, still chewing, made something like a bow, disappeared, and slammed the door behind him.

The adjutant and doctor sighed loudly. Then they remembered that Reisiger was sitting with them, laughed, and cleared up. The adjutant, probably only a little older than Reisiger, offered him a cigarette. "You needn't be frightened, Herr Kamerad. You'll have to get used to that tone. It's not that bad, by the way. Brett's a fine fellow..."

The doctor interjected, "It's just a pity he so rarely lets himself be noticed."

What can one say to that? Reisiger smiled. Then he said, "By the way, I've never been an ordnance officer in my life. Perhaps you would be kind enough to show me a little of what I have to do."

The adjutant: "With pleasure. That said, it's not easy to list everything. The daily routine will be roughly as follows: eight o'clock in the morning, going through weather reports for all batteries. 9:15 a.m. issue of orders to all batteries. Ten to 2:00 p.m. ride through the positions with the captain. 2:30 p.m. report on manpower. 2:40 p.m. report on the number of horses. 3:10 p.m. passing on all news from the enemy. Four o'clock telephone consultation with all battery commanders about their wishes. From 5:00 to 7:00 p.m. work with the captain, new firing regulations and other technical questions. Around eight o'clock in the evening, procurement of material for the expansion of the batteries. Issue of orders around nine o'clock. Around 10:00 p.m. receipt of divisional and regimental orders. Yes . . . and then the captain usually either plays cards until twelve or one o'clock or takes one of us to the nearest casino and drinks there . . . . Besides, you have to be prepared for someone wanting something from you on the telephone at least twenty times a night . . ." He laughed. "Well, you'll see. In any case, there's more to do than we'd all like."

"Why," Reisiger asked, "did the captain actually fall for me?"

The adjutant seemed to squirm a little. Finally, he said, "Well, Herr Kamerad, if I'm being completely honest, it's because you don't have a good standing with the regiment. The captain doesn't like tame officers. He must, if I may say so, always have gentlemen with whom something is wrong."

A strange explanation. Reisiger was stunned. He didn't quite know whether he should be depressed or proud of his appointment. In the end, he opted for pride. *As far as I'm concerned,* he thought, *I don't know how I earned it. But if people think I'm a felon, this is probably the best way to put up with it here on the staff.*

7

Reisiger got on well with the captain. He shouted at him about two dozen times a day in a way that you certainly shouldn't shout at a recruit these days, but that didn't spoil the good relationship at all. In fact, a miracle happened several times that Weller had previously thought impossible: Brett received lard or butter from home almost every day, and it had already happened twice during morning coffee that he freely smeared a thimbleful of fat on his ordnance officer's saucer with the broken tip of his penknife.

The relationship between Reisiger and Weller was good, very good. Sometimes at night after work, when for once the captain had spared them both from drinking in the officers' mess, they would sit together, forgetting war and uniform, discussing, reading to each other. Or got up to mischief. Weller had a gramophone. Reisiger had taken the spring out of the case and put it back in upside down. A wonderful pleasure. Many a lonely hour was spent playing the five available records, from

"Lieben Augustin" to "Tannhäuser" from back to front; always accompanied by alcohol, plenty of cherry brandy from Weller's homeland. It was bearable. And when the doctor joined them, a student like the other two, they celebrated orgies of mischief.

Now it was February. *February?* Carnival! And several times the three of them had a regular masquerade party. They appeared in pajamas, with their uniform skirts turned upside down, paper caps on their heads, painted with ink, roaring and dancing.

Wonderful mischief that made Reisiger forget a little of the desolation that he sometimes sank into.

And wasn't this desolation justified?

Nonsensical work! Phone call after phone call. Message after message. Writing lists, filling files, reports, vacation approvals. Files, files, paper, paper.

Nothing about the war, nothing about the military (except on paper).

Nothing about comrades and soldiers (except on paper).

It got worse and worse. It got suspiciously bad: "It must be reported immediately how many horses . . ." "Report immediately how many officers and men . . ." "Report immediately how much ammunition . . ."

Ammunition?

It's a ceasefire.

"Report immediately how much ammunition, separated into shells and shrapnel . . ."

And a new telegram from the regiment: "All batteries of the Third Battalion will receive 300 pieces of shrapnel per gun tonight."

Tonight? It is already two o'clock.

So far, Reisiger has had to write stock lists for the number of horses. Dead tired. He has the battery commanders woken up, called to the telephone, and gives the order to each one personally. He hears some astonishment, some oohs, some whys.

He sinks into bed around three o'clock.

After a short time, someone bangs on his bedroom door and shouts. Reisiger turns on the light. The door flies at his feet. Standing in front of him, swaying, capless, is the captain, a bottle of champagne in each hand. He laughs and shouts at him: "Come on, get up, Reisiger. Lieutenant Karl from the Eighth Battery is here too. We'll let the Weller sleep. You're our third man at skat!" Then his laughter becomes raw and loud. "Come on Reisiger, I've brought something too. Here ... the army report. Here it is. So come on."

And when Reisiger appears distraught, quivering with rage, half-dressed, the captain lies down across the table and reads from a message pad, bellowing: "Here ... Eastern theater of war ... that's us, you understand, Reisiger ... don't look at me so stupidly.... What does it say here ... so pay attention ... Well, the bastard is lying on his head ... so, so here, 'The armistice expires on February 18, twelve noon.' Signature ... well, I wonder who ... Reisiger, you're too stupid ... it says clearly, and I'm telling you, that's my man, understood: Ludendorff.[19]"

Armistice expires? "So, expires. - Zero ouvert - armistice expires - aha, so the Russians are no longer allowed in our trenches - thank you, I'll pass - expires, is over, is finished. Yes, Herr Hauptmann give - yes, thank you - armistice - so the killing begins again - we are being led back to glorious times - Ludendorff - No, Herr Hauptmann, I have nothing - I pass - expires - Coeur is trump - armistice expires, is over."

Reisiger is allowed to go to bed for a short hour at half-past six in the morning.

"You've lost big, Reisiger," slurs the captain.

8

At 10:00 a.m. on February 18, 1918, the detachments of Field Artillery Regiment 253 are ordered to ensure that all batteries are made ready to fire, all guns ready to sound the alarm. The order to advance can still come today.

The regimental adjutant is on the line: "Herr Leutnant Reisiger? Please compare the exact time: It is ten o'clock seven minutes .... You're still receiving written orders from the dispatch rider ... Will you please ensure that all the batteries in your division open fire on the

---

[19] Erich Ludendorff. 1865–1937. Prussian general. Responsible for Germany's military policy and strategy in a de facto military dictatorship in the latter years of World War I. Serving as chief quartermaster general of the German Army, Ludendorff was the architect of the operation to break the stalemate along the Western Front in Spring 1918. Participated in Hitler's attempted coup d'état in 1923.

Russian positions at twelve noon sharp .... The regimental commander emphasizes that the fire attack is not just a demonstration but asks you to make sure that the enemy trench is taken under effective fire. Each battery has eighty rounds at its disposal."

Reisiger rings the telephone switchboard. "Please put all battery commanders on the line."

After a while: "This is battery commanders 7th, 8th, 9th Battery 253."

"Good afternoon, gentlemen ... please compare ... exact time ... all batteries are firing at twelve o'clock sharp."

A few minutes before twelve, Captain Brett is standing with his adjutant and ordnance officer on the village street in front of the staff quarters. Soldiers have gathered in front of all the houses.

Nobody is talking.

Twelve o'clock. Thunder at the front, as far as you can hear. The captain strikes the gaiters with his riding whip. "Well, thank God, the war is finally back. We have our honor back. Gentlemen, I'll buy you a bottle of champagne."

As Reisiger goes into the house, he can feel that he's swallowing tears. *Yes, yes, now we have our honor back.* He sits down silently at the table. He looks at Weller, who is silently playing with a pencil. He tries to imagine how he would have felt if he'd had to be a gunner earlier. *I think, he thinks, I would have gone on strike.* He bites his lower lip. *What does strike mean? The men must feel the same way I do: That it's meant to suddenly start shooting again. And yet they have to do it. We all must do what we're told.*

He suddenly addresses Weller sharply: "Tell me, who actually gives orders like that?"

Weller shrugs his shoulders and carries on playing.

Reisiger thinks of New Year's Eve 1915, how suddenly, out of the blue, they had to bang into the night without any reason .... He looks at the wall and says, "We're going to kill ourselves with our grit."

Here comes Brett, swinging the champagne bottles. "Urban, quickly, the glasses. Gentlemen, long live His Majesty and the war!"

9

Regiment 253 had advanced a few days after the armistice expired. Ready to fire against an enemy that was not there. What had the war become? Here in the East? Advancing against an enemy who no longer existed, who had preferred not to be an enemy, who now lay in the villages without uniform, exhausted to the point of exhaustion, defenseless. The Germans? Very well, let them come. Russia is big. Let them march. Enemy? We are not an enemy. You won't find an enemy.

So Regiment 253 marches on and on, day after day, into Russia in a large detachment.

What use are cannons, what use are officers, what use are soldiers if there is no enemy?

For the first few days, all this is taken with mysterious importance. The ears are pricked up, the senses are tense. There's a forest; is the enemy there? Guns are loaded and secured, in front of each battery is a vanguard squad, sent into the undergrowth. Day marches without paths, still deep in the snow, still very cold, through the endless forest. Bivouac at night with small fires. Patrols, guard posts: if only no enemy takes us by surprise.

Russia is big. There is no enemy.

The atmosphere is getting dangerous. *Jesus, what are we doing here, soldiers, officers, cannons without an enemy?*

Further into Russia. There is an enormous amount to eat, because you can shoot mutton or fat pigs in every village you reach after many hours of marching. But that's the soldier's job: shooting mutton and fat pigs?

The teams get tired. The horses can hardly pull. The weather is so bad that the guns threaten to rust. Forward, forward! And among the officers the restlessness: If only something like the enemy would come!

At last, thank God, the staff riding ahead of the division comes across a burnt, ruined palace in a village. Heaven be praised; the enemy must be here.

Interpreter approach! The enemy? The villagers laugh with reassurance and amusement. Yes, of course, it was the Bolsheviks.

Finally, the German advance has a sufficient slogan: "Watch out, we're about to meet the Red Army. Nobody knows what that is. But it sounds good, mysterious, and desirable. It makes the soldiers ready to fight again, takes the rust off the cannons, and gives the officers a dashing backbone. Ha, Red Army!

After forty-eight hours, another burnt palace; after two days of marching, still a burnt ruin. Again and again, they shoot at mutton and fat pigs, eat, and listen with greed to rumors that come from far away Russia: Once a Red Army troop wiped out a German detachment. Once a German detachment unceremoniously put some Bolsheviks up against the wall.

There is almost something like enthusiasm again: Hurray, long live the war.

But this enthusiasm doesn't last, because it always stays with the mutton and fat pigs. The soldiers are tired again; the guns are rusting; the officers are bored.

There is a strange saying going around Russia!

"In the West, a brave army fights a war,

in the East, the fire brigade fills their jar."

The soldiers' mouths grin as they sing these beautiful verses. The song is sung day and night.

The detachment with III/253 stands on the banks of the Düna River one morning. Conquered without a fight. The officers grumble, the men grumble, the guns are silent, and mutton and fat pigs seem to have died out here. There's only one thing left that can save us. It arrives! An order: The Reserve Field Artillery Regiment 253 is to be deployed to the west.

10

Gentlemen, the first time I had the honor of addressing the Reichstag in this place, on November 29 of last year, I was able to inform the House that the Russian government had sent a proposal to all the belligerent powers to enter into negotiations for an armistice and a general peace. We and our allies accepted the proposal and immediately sent delegates to Brest-Litovsk. The powers hitherto allied with Russia did not accept the invitation.

The course of the negotiations is known to the gentlemen. . . They remember the repeated interruptions, the breaking off and resumption of the negotiations. They had reached a point where an either/or had to be spoken.

On March 3, the peace treaty was signed in Brest-Litovsk, and on the 16th of that month it was ratified in Moscow by the competent assembly. . .

But, gentlemen, we must not deceive ourselves: World peace is not yet here! Unfortunately, there is still not the slightest inclination on the part of the Entente states to refrain from the terrible craft of war; they still seem to be pursuing the will to continue the war to the utmost, to the point of our annihilation. We will not lose heart over this. We are prepared for anything. We are prepared to make further heavy sacrifices.

God, who has helped us so far, will continue to help us. We trust in our righteous cause; we trust in our incomparable army, its glorious leaders, its heroic fighters; we trust in our brave, steadfast people. But the responsibility, gentlemen, for all the bloodshed will fall on the heads of those who, in frivolous obduracy, do not give ear to the voice of peace.

(Reich Chancellor Count von Hertling, Reichstag,
March 18, 1918)

Chapter V

1

On the journey from Russia to France, Regiment 253 lost seven officers ... They all fell ill, they said, with stomach ailments. No doctor could cure them.

2

The transports had a seven-hour layover in Berlin. The crews were forbidden to leave the train station. The officers were given city leave.

A leaflet was distributed on the platforms.

Berlin, November 1, 1917

A word to the Herren Comrades!

Maintaining discipline in a city like Berlin at a time like the one we are living through is a serious task. I ask each and every one of you: "Help me to solve this task." The army has become large and strong through discipline, and we have it to thank in good part for the successes we have achieved so far against numerically superior enemies. It is the duty of all of us to maintain and promote this manly discipline among the troops here, at home as well, in order to remain victorious in the great battle, which the majority of the soldiers here are still or again called upon to carry out to a happy end. The good example set by the officers in this respect will be most conducive to this high purpose. Even if the one or other regulation or institution made especially for Berlin may seem annoying or superfluous to the individual coming from the field, be convinced that these regulations have a specific reason and are only intended to serve the good of the whole.

Exercise self-restraint in your behavior on the street, in restaurants, and in your dealings with each other and with subordinates. Rebuke officers who violate our good professional customs in a comradely manner and report them to me, regardless of the person, for offenses that affect our professional honor.

My recent experiences have prompted me to draw particular attention to the following points:

Open paletot, walking stick—unless the wound urgently requires one—carrying a weapon without gloves is not compatible with the ideas of proper street dress. Letting oneself go in this respect is only too quickly imitated by ordinary soldiers.

It is unmanly and improper for a healthy officer to seek support from a lady and allow himself to be led by her.

It is not customary for officers to keep one hand in their paletto pocket when greeting their comrades, or to keep their cap on their head when entering a public place. . . If the gentlemen have taken window seats in larger cafés, it is improper for them to take no notice of passing comrades.

I particularly warn the younger officers, where they have the opportunity to intervene on the street or in pubs, against an overly brusque tone, which is likely to cause displeasure among the public. It should be borne in mind that there are old conscripts and young men among the people who have voluntarily rushed to the flag and still lack a thorough military education. A serious admonition, not without a certain comradely benevolence, will often be better.

The unanimous enthusiasm, the absolute will to persevere in unshakeable confidence of victory, the willingness to make sacrifices in all classes of our people can easily falter if the officer, wherever he appears, does not maintain the attitude that corresponds to the seriousness of our time. The prestige of our profession, which has been sealed by the brilliant successes of our weapons on the battlefields, even in those classes of the people who looked with disfavor on the officer before the war, is a great asset that everyone must strive to preserve. It is, therefore, of the utmost importance that the officer presents himself everywhere with the dignity befitting our rank. It is a very regrettable fact that misconduct on the part of individuals has unfortunately caused righteous indignation in wide circles of the population on several occasions, which compels me to conclude by emphasizing the points that have given rise to this.

It is unworthy of an officer to show himself in uniform with dubious women in the street or in pubs; he lacks tact if he visits pubs in such company, where he can expect to meet comrades with their ladies. Entering cabarets, bars, night cafés, and similar establishments in uniform is a denigration of our dress of honor.

The officer must beware of excessive consumption of alcohol, which can only delay the healing of a wound, just as he must be careful in his whole way of life to spare his body in order to stand his ground again before the enemy. Whoever does not keep this in mind is sinning against himself and against the Fatherland, which today, more than ever, urgently needs healthy officers at the front.

The Commander
v. Bonin, Lieutenant General

3

No weapon, no gloves, don't get along ... when the healthy officer seeks support from a lady ... when one hand remains sunk in the paletto pocket ... which corresponds to the seriousness of our time ... even in those sections of the population who looked with disfavor on the officer before the war ...

4

I regard Social Democracy as a passing phenomenon; it will pass.
(Wilhelm II., January 9, 1900)

5

After the long journey, F.A.R. 253 arrived at a military training area near Valenciennes.

For the first time, they came into contact with many new means of combat, which they had only heard rumors of in Russia. Gas ammunition, anti-tank shells. What for?

There were drills. And courses were held for the officers of all staffs and batteries.

After every lecture it became clear: The battle of the machines would always be more decisive, and the artillery would have the last word. People relearned. New lessons were learned. All the old regulations were replaced. Every officer walked around with books under his arm, training regulations, firing instructions, combat paragraphs.

An unpleasant atmosphere was spreading. It was all too unpleasant, people sensed—

Added to this was the lack of good officers, the absence due to sick leave during the transport. It looked like an appointment, like shirking. People didn't want to believe it. But they mocked it. And they let it depress them. Should the stay in Russia have been enough to demoralize them? If the men noticed?

And, above all, where would they get new officers from?

The III Battalion 253 needed four. Inquiry at the replacement department in the garrison: no one fit for war service was available.

Discussions: Four sergeants were promoted within a few days. *But what could be done with these young officers?* None of them even knew the West. It was very questionable how they would behave.

New briefing: Who from the regiment had even been to France?

The report from the III Division did not give a good result: almost 99 percent of all the men had been in Russia since 1914. Likewise, 80 percent of the officers. None of the staff was familiar with conditions in the West, except Reisiger.

6

The situation in Russia and Italy will probably make it possible to strike a blow in the Western theater of war in the new year. The balance of forces on both sides will be roughly equal. About thirty-five divisions and 1,000 heavy guns can be made available for an offensive. They will be sufficient for one offensive; a second, larger simultaneous offensive, for example, as a diversion, will not be possible.

Our overall situation demands that we strike as early as possible, preferably at the end of February or beginning of March, before the Americans can throw strong forces into the fray.
We must beat the British.
The operations are to be based on these three guiding principles.

(Guidelines of the Supreme Army Command for the preparations of the spring offensive 1918. Discussion between Ludendorff and the army group commanders, Mons, November 11, 1917)

7

WTB reports June 6, 1918: Destruction of the proud maneuvering army of the Entente. American troops in the West have grown from 200,000 at the beginning of 1918 to 700,000.

(Leop. Schwarzschild, diary, November 3, 1918)

8

The great battle raged at the front. The army reports were followed day by day. Was progress being made? Was the war of movements finally coming, longed for by everyone? Would it bring the final end?

The army reports were followed. But the phrases did not make the people any happier or braver. The regimental commander often summoned his officers in the evening, and then the German attack was discussed on the basis of maps, as far as documents could be found. Then one June night: "Gentlemen, as far as I can see, a lot has been achieved in the last few months, but Reims is missing.... Our attack seems to be over for the time being. But remember the words of General Ludendorff: We must not believe that we will have an offensive, as in Galicia or in Italy; it will be a tremendous struggle that will begin in one place, continue in another, and take a long time; it will be difficult, but it will be victorious.... Gentlemen, I wish us all that our regiment will soon be part of it."

Two days later, march off! All the orders that arrived were excitingly mysterious. We knew nothing, learned nothing, could hardly think or discuss more than a few hours of marching. Three nights, always twelve hours on horseback, through villages and towns. The formations were torn apart.

The staff of the III Battalion rode alone, accompanied only by a few dispatch riders from the three subordinate batteries. And always after three or four hours, an officer stood by the road, an ordnance officer of the division, with a sealed letter in his hand: destination for the next few hours.

One morning: Laon. There was to be a long rest here. The batteries were accommodated in small villages near the town or in the large forest that was everywhere.

There was no sign of the front. We were closer to it; there was incessant thunder in the air. Despite the defensive batteries, flocks of enemy planes appeared in the sky, bombing Laon and the villages. But otherwise, there was peace.

The men took advantage of it.

On the slope of a hill stood some wooden huts, perhaps built months ago by the infantry. They became quarters for the division staff.

There was nothing to do. Not even telephone lines were laid to the regimental headquarters. An order came every evening, and every evening, we trembled in anticipation of its contents: When are we going forward? But there was silence. More than that, explicit orders were given to give the horses and crews unlimited rest if possible. There was also plenty of food, lots of bread, lots of dried fish. Sometimes, a quarter of a liter of brandy per head.

Weller and Reisiger played chess most of the day. The captain had discovered a very good canteen belonging to a cavalry regiment in the area. He actually always stayed there.

He was also there one evening shortly before dark when Weller and Reisiger reached him to read him the order that put an end to the peace.

Rear Inspection VII, Ia, 365 secret.

According to A.O.K. 7, Ia Gen. Art. 4431/18 of June 26, 1918, the Longueval area is to be reached in two night marches, starting in the evening of July 1, 1918. March a) L.M.K. on the night of July 1 in the evening b) Regimental and Divisional Staffs with commanders, as well as an officer and escort from each battery in the evening of the night of July 1. c) Batteries without large baggage in the evening of the night of July 2. Marching route Presless, Lierval, Grandelain, Bray-En-Laonnois, Soupir, Chavonne, St. Mard, Vauxtin.

Marching distance every night is about twenty kilometers.

Direction of march Fismes, where further marching quarters are to be requested from the local commandant's office by an officer to be sent ahead on each previous day. A special officer from the A.O.K. will provide the necessary information there.

In principle, only night marches may be carried out and in such a way that the march is only carried out when darkness falls, and the new bivouacs are already reached at the beginning of dawn.

In all formations, it must be pointed out with the utmost strictness that when marching at night, all vehicles, etc., must be kept under roadside trees as far as possible when enemy aircraft are approaching, and that all movement on the open road must be stopped immediately. And that, furthermore, every bivouac should be kept out of enemy sight during the day. Horses and vehicles must be sheltered under trees and the remains of houses; even in villages, cover must be taken against sight when enemy aircraft are approaching. Campfires are forbidden. Smoke must be avoided. Accommodation is not to be expected.

All formations have to provide themselves with rations at the Athis or Laon stage magazines in such a way that they are still equipped with rations for two days after crossing the Piermand-Laon-Sissonne line.

Do not take large bags with you! Please note that these measures will be monitored. Take food and fodder wagons with you, as well as water wagons. Attention is drawn to the poor water supply on the march via Chemin des Dames.

Addition of the Regiment: The route and destination may only be announced to officers. Instruct crews in detail about behavior on the march, in bivouac and in case of danger from aircraft. Be careful with flashlights and lighting matches on the march!

Cover epaulets and numbers on vehicles.

Captain Brett read the order, pushed it back to Weller: "Gentlemen, I think this is the final death sentence."

9

"Not to count on shelter."

"We draw your attention to the poor water supply on the march via Chemin des Dames."

The regiment marches.

In front, the squads and the commanders with their staff.

Reisiger next to Weller every night.

Almost always at a trot. Toward morning the troop splits up, the section commanders have their batteries brought forward and move into bivouac.

The most desolate, horrible area Reisiger had ever seen.

Impossible to talk when riding next to Weller. Wherever the horse goes, wherever the gaze falls, it is nothing but the final destruction. The vehicles are ordered to be placed under trees if possible. There is nothing but the last remnants of splintered and shattered trunks. They are to take cover under the remains of houses; there is nothing but lime fields, dusted into atoms.

There are signs on them, signs again and again: "Here was the village X . . . Here was the village Z."

The nights are dark. It is all the more terrifying toward dawn when the veils are lifted. The most deserted, coldest, most horrific crater landscape. No grass, no flower, no stone next to or on top of another. Nothing but deep holes, some filled with greenish, stinking water. Unspeakable exertion. Often, for half an hour, you get off your horse, take the animal by the halter, and stumble along in the prescribed direction. Often toward morning, instead of the expected battery of four guns, only one gun arrives. The others have broken down, somewhere in the dirt. It may take hours for them to follow.

No water except in the meager barrels they carry. No one can wash themselves. After eight days on the march, the men and officers are thickly encrusted and have beards.

As they were not allowed to cook, they subsisted on bread and sandwiches and cold canned meat.

And every time you hope for an end, a goal, the officer reappears with the sealed order. And you sneak on. It is inconceivable that this troop, which gives the impression that it consists only of sick people, should go into battle.

After ten days, the area improves. Now trees are growing; now there are the remains of houses; now there is a village. The officer is standing there again. The order is unsealed: The staff of the III Battalion moves into quarters here in the village. The next night, the batteries move to the designated firing positions. And the front? "Yes, the front is near."

It is strangely quiet. Not a single shot is fired during the day.

Now, toward evening, you rarely hear a small-caliber shell.

The captain is sitting with his two officers and Winkel in a large room in the staff quarters. The three battery commanders come in to report to the firing position.

With questioning looks: What's going to happen?

Nobody knows anything. Not even the captain. The regimental order has arrived in the meantime, with trivial details. The captain shakes the officers' hands jovially: "But gentlemen, what's going to happen? We've certainly been lucky again. All this cinnabar seems to be taking place in another region. You've heard; complete silence at the front." He interrupts himself. An ordnance officer from the regiment has entered. "New order for the III Battalion."

The captain opens the letter. "Children, are you nervous? You're driving me crazy with your secrecy. What's going on?"

The ordnance officer shrugs his shoulders. "I don't know anything either, Herr Hauptmann."

Order, secret, to be written and transmitted only by officer: The batteries will be deployed as follows: from ten o'clock this evening, only one gun from each battery will be deployed every hour. After one hour the next and so on. The utmost care must be taken to ensure that the deployment is completely silent. Wheels of gun carriages and gun mounts must be wrapped in straw. The crews must be informed that any use of flashlights or matches and smoking will be severely punished!

Under no circumstances may telephones be used. Any laying of lines is forbidden!

Instead, an officer from each battery will stay day and night with a horse at his battalion staff—all positions are already amply supplied with ammunition. As soon as the last gun of each battery has been posted, all officers and men will leave the position except for a guard of two men each, who are to be warned under threat of severe punishment that under no circumstances may even the slightest movement be shown within the position at daybreak.

The officers and men who have been withdrawn will lie in the caves indicated on the attached sketch about three kilometers behind their batteries until further notice. For them, too, complete immobility applies from morning to night.

From three o'clock in the morning, no teams or riders are allowed to pass through the area. Rations are brought to the position and the standby area at night by pedestrians. No mail is to be expected in the next few days. It is also forbidden to send field post.

The captain places the order on the table. The regimental ordnance officer bows and leaves. The captain: "Everything all right, gentlemen? Don't you understand, everything that happens is so secret under all circumstances that even the crews are kept absolutely in the dark? Send me your youngest soldier as messaging officer . . . . Thank you very much. Shot in the throat and stomach!"

The regimental ordnance officer comes three more times that night. The captain has long since gone to bed, which is on the top floor of the house. Upstairs, in the room next to him, the battery officers are also lying down.

Weller and Reisiger are downstairs alone with a few scribes and draughtsmen, working on the new orders.

Sleep is out of the question. You can feel that a machine is being set in motion here, which, despite the enormous scale of the performance it demands, must have the most difficult precision as its most important prerequisite.

The accuracy of the number alone can guarantee this precision! And the frenzy of numbers is so great that Weller and Reisiger have long since forgotten how each number means one, one hundred, one hundred thousand people; how a millimeter on the maps in front of them determines the death of a battery; how a red ink circle commands a hail of gas shells, a black square a rain of shrapnel.

New maps, aerial photographs, reconnaissance reports come with new dispatch riders. The picture of the front becomes clearer and clearer, defined for every foot width. And finally, apart from about forty kilometers to the right and left of Reims, the longed-for, longed-for goal, there was not a single trench, not a single trench, not a single ready position, not a single light or heavy battery of the enemy that was not marked with meter precision on the increasingly perfected map formations.

Each enemy gun, each ridiculously insignificant position, is given a number.

Each of these hundred numbers is distributed among the concentrated mass of German artillery.

Mountains of ammunition have been lying ready for each number for days.

But will it all work? When the big order comes, will every single shot really hit where it is needed?

Nights pass over the calculations.

First comes the work, numbers, numbers, measurements, this encircling of the enemy, this circling around: here must be a hundred shots, here twenty, here eighty, here must be drummed—it all comes flooding back again and again as a frenzy in Reisiger. It's always a new frenzy.

"Look, here, Weller, behind the depression in the ground, there's an excellently covered battery, as the new aerial picture clearly shows. Which we are to flank. What do you think . . . three hundred, well, five hundred rounds on it? Our Eighth Battery can still do that. Can't it? All right, then. Make a note, on target 318e, 8/253 of X plus 80 minutes will fire 500 rounds of bunt."

But then, when the two officers have presented the lists and plans to the captain in the morning, when everything has been calculated and noted down for the third time, when the two are allowed to sleep, then the intoxication is chased away.

Then comes paralyzing exhaustion, so great that sleep is out of the question.

Then they toss and turn, on one side, on the other. The images on the maps tremble in front of their burning, convulsively closed eyes. "Jesus, Reisiger, haven't we forgotten the crossroads in square 17 above trench 107?"

"Tell me, Weller, I'm afraid our Ninth Battery doesn't yet have the exact details for the position of Battery 33 Caesar."

"What, Reisiger, is the 7th doing at X plus twenty minutes?"

And, all confused, finally dragged into half-sleep. "X plus twenty minutes? X plus, what did you mean, Weller—X plus twenty minutes, X plus X, X . . ."

And awake again, upright from the mattress: *X, X? What will X be? When is X? What day is X? What hour is X?*

Looking at the clock: two o'clock at noon. Drop your hand: Is the great hour X in twelve hours?

And when this great hour strikes X, will everything really be all right? Will the machine not lose synchronization for one second, from X to X plus sixty minutes and again plus sixty minutes and up to X plus infinity?

Get up. Aha, Weller is really asleep. Chewing some dry bread. "Good afternoon, Herr Hauptmann." He's now sitting over the maps, leafing through piles of files, reading the new combat regulations, leafing through them like you're looking for passages in the Bible, a few lines here, a few lines there. "Reisiger, this is the best book there is. Excellently written."

Reisiger already knows the phrase. The old fad since the training ground. Dad with his family, reading aloud. "I'll have to take care of the weather reports, Herr Hauptmann."

The captain turns the pages faster. "Listen to me first. Pay attention, here's what it says, here, paragraph 53:

'*Complete and continuous defeat of enemy artillery is rarely successful. Depending on the combat situation, a distinction must, therefore, be made as to whether the aim is to paralyze the enemy artillery for a short time or at least to dampen it, or whether the destruction of its equipment, ammunition, and shelters is intended.*

In the former case, gas ammunition is particularly effective. Firing with brisance ammunition only promises success if the amount of ammunition used is not too high, if it can be fired under visual observation or if it is at least possible to keep the scattering limits within narrow limits. Whether to use sudden destructive fire or slow disruptive fire depends on the combat and ammunition situation.

If, on the other hand, destruction of equipment, etc., is intended, the enemy artillery is to be engaged with carefully directed destructive fire until the individual target in question has been destroyed. In addition to the use of sufficient ammunition, success depends largely on visual observation, which should be carried out until the end of the firing. If there is no visual observation, there must at least be a sufficiently detailed firing plan in place.'

"Well, thank God we have that...." He continues turning the pages, licking his finger. Reisiger wants to go to the door. "Just a moment, wait a minute ... look here, for our situation, mind you: paragraph 54, down here:

'The artillery preparation of the own infantry attack usually takes place in calm destructive fire until there is a probability that the attack targets are well and evenly under fire. Recognized, particularly important targets (melee guns, machine guns, dugouts, etc.) must have been destroyed to a greater extent.'"

Weller looks up. "This is guaranteed to have happened completely at X o'clock plus ninety minutes." Then he continues with a grin:

"'The fire then increases to annihilation fire, which, however, must still alternate with destructive fire to protect the equipment. The infantry attack is accompanied by the destructive fire until it breaks into the enemy line. Then the artillery transfers the fire in leaps forward. The further procedure depends on the military intentions and must, therefore, be precisely regulated!'"

He closes the notebook, laughs, and says, "The further procedure depends on your intentions, ha, well, Reisiger, we're making the race."

That was on July 13, 1918.

10

As early as July 7, the leader of the 4th French Army, against which the attack was primarily directed, was able to tell his troops:
"You all know that there will never be a defensive battle under more favorable conditions. We are prepared and on our guard. You are fighting

in the terrain which you have turned into a mighty fortress by your labor and steadfastness."

(The Great War, 1914–1918, Concise account based on the official sources of the Reich Archives by Erich Otto Volkmann, Berlin, 1922, p. 178)

11

The intelligence officer of Army Group Rupprecht (of Bavaria) and, independently of him, I as the intelligence officer of Army Group Crown Prince, were able to report early on, on the basis of investigations from prisoner statements, captured written material, espionage and finally even defector treachery, when approximately and where exactly the first Foch counter-offensive would take place.

On July 11, exactly eight days before the big turning point, I sent a detailed report to the Supreme Army Command and all higher command posts on the Western Front about the assembly of about twenty of the best French combat and assault divisions in the Villers Cotterêts forest and west of it, which had been taking place for three weeks. Each French division and each of the five American divisions involved could be identified by number and, according to the defector's statement, Foch's combat reserve was to attack the very weakly occupied German front "around July 17" with the right wing on the Ourqc and the left wing towards Soissons, twenty kilometers wide.

Particularly fortunate circumstances and coincidences had thus made it possible this time to have prior knowledge of the enemy's intentions, which is otherwise extremely rare in war.

But this report and similar ones from other intelligence officers of the Western Front armies made little impression on the German Supreme Command. The reason for this is perhaps clear from a telephone conversation I had on July 13 with a senior intelligence officer at the Grand Headquarters. I asked him anxiously and uneasily whether nothing further would be done in response to my report of July 11, meaning above all a reinforcement of the threatened section of the front. The forces for this were available, as there were still around twenty German combat divisions on standby in Belgium at that time for a major German attack planned there against the British front. I had believed that at least a few divisions would now be sent from there behind our threatened section of the front, which was occupied by only nine divisions over a width of about twenty kilometers, and these were severely weakened. However, my nervous questions were dismissed by the Supreme Army Command with the brief reply: "Just let Foch assemble what he wants and plan what he wants. Before he makes the attack that you have so beautifully scented, we will dictate to him where he has to fight, as a defender, not as an attacker!"

(Major (retired) Kurt Anker, *Berliner Tageblatt*, July 20, 1929)

12

On July 13, 1918, reports came from the front that some of the troops were suffering from severe influenza. Orders: everything stays in position if possible!

Influenza, usually epidemic, very contagious infectious disease caused by the influenza bacterium. (Der Kleine Brockhaus, 1915)

On July 13, 1918, there were reports of severe attacks of dysentery at some points on the front. Orders: The medical stations and field hospitals are to provide accommodation for the sick. If possible, care must be taken to ensure that the troops are brought back to normal strength as quickly as possible.

Dysentery, epidemic, diphtheritic inflammation of the mucous membranes of the large intestine; fever, abdominal pain, excruciating urge to defecate, and diarrhea, with mucous or bloody stools being defecated with great pain; can become chronic, but can also be fatal due to exhaustion. (Der Kleine Brockhaus, 1915)

Lieutenant Reisiger is sitting next to Lieutenant Weller at about ten o'clock in the evening, bent over maps and lists. So bent that his chin is pressed against the edge of the table. He can hardly speak, has a red haze in front of his eyes, and has fire in his body. He is in pain, groaning out every number, every detail.

The doctor, Winkel, comes into the room. He sees Reisiger and touches his forehead: "You have a fever."

Reisiger: "That's a feat ... I've only been shitting pure blood since noon today."

Winkel: "But Herr Reisiger, you belong in the military hospital. You have dysentery."

Weller: "In plan square 18, small square 8 is slash 11 ..."

Reisiger, roaring: "Doctor, this is maddening."

Weller: "We'll be done soon, Reisiger ..."

Reisiger jumps up, rushes out the door. Returns after a while: "I've got nothing left in me at all. It's pushing so terribly in my intestines."

The doctor feels his pulse: "Herr Weller, I'll report to the captain that Herr Reisiger is absolutely ill ..."

The regimental ordnance officer enters the room. "Morning, gentlemen. Please see to it that all the officers of the battalion report to the divisional headquarters in the church in the village at three o'clock tonight. Plus, of course, the battalion commander and all of you. "Now it's ..." he looks at his watch, "three quarters of twelve. So, have a good time. An hour's ride at most." He scratches his temple and says, "Gentlemen, the place is buzzing, I think. Maybe we'll get going ... Tomorrow!"

Weller, out the door: "Saddle the horses."

Reisiger, up to the captain, who is fast asleep: "Herr Hauptmann, we have to go to the division tonight."

The doctor: "But, Herr Reisiger, you mustn't. It's pure madness. You'll fall off your horse."

Reisiger, tired, waving his hand: "Shot to death or shit to death, dear doctor ... it doesn't matter."

13

Three o'clock in the morning, in the night of July 13 to July 14, the church, in the village where the staff is located, is packed with artillery officers of all ranks. In all the galleries, all the aisles, in the nave, officer after officer. Majors and the youngest lieutenants.

The room is dimly lit with candles. At the altar are large black candelabras with burning bundles of candles. The altarpiece, Jesus on the cross, touchingly painted, is dark. Only the strangely chalky face of Christ is visible. Two badly drawn, very long-fingered, white, pierced hands.

For a few seconds, the congregation has forgotten that they are not gathered in a church, but at a war council; there is only very quiet whispering.

The church bell chimes three.

The field constabulary tear open the church gate. A small officer with a hooked nose appears with echoing steps. A blue file cover under his arm. He keeps his cap on his head, takes no notice of the congregation, and walks to the altar steps.

His back against the image of Christ.

It is the advisor of the Army Group in artillery matters.

He opens the file cover and begins to speak. In a cutting voice, cold, matter-of-fact, to the point. Says that they have gotten out of the habit of making any predictions. You must stand on the ground of facts, to the utmost. Raises his voice a little: "Gentlemen, it doesn't matter whether what we are planning brings the final decision or not. I hope you have all done what is humanly possible. I wish, and I convey the clear will of the Supreme Army Command, that you have done more. We leave phrases to the enemy. We act. We will win. Or we die.

"And now to the point: Gentlemen, we are working, or rather you are working, for the first time in this campaign with two unknowns: No one knows on what day and at what hour our operation will begin. Under no circumstances should this lead to laxity. On the contrary, gentlemen, we expect you to be ready at any moment, and that this readiness, even if it lasts for days, must not become lax.

"The name of the undertaking is Anna. The time is X. We don't need to worry about anything else. On the cue Anna, all positions will be occupied. If this cue is followed by the order, X is eleven o'clock . . . or what do I know . . . the guns will start their predetermined fire at the time indicated.

"That must work. And gentlemen, everything else will take care of itself.

"We are not here to determine whether the time is serious, whether the enemy is strong or weak. We are only here to act.

"And, gentlemen, this is important, to act in silence. Our undertaking promises success for all. But it is absolutely essential that the enemy does not receive any news of it."

The major closes his cardboard cover and puts his hand on his cap. Those present bow. The major walks stiffly down the center aisle.

The church empties.

14

The two unknowns, the date and hour of the "Anna" undertaking, now become a torment.

You have been working until now, but now that everything is on paper down to the millimeter and the second, when there is nothing left to do but wait, normal thinking turns into the senseless spinning of a merry-go-round. You are alone, or you are with others, you sit and stare in front of you, or you try to have a conversation, you chew on the horribly dry, crumbly bread or you smoke: It's always circling senselessly to senseless tunes, whining, hissing—Anna, X, X, Anna.

This begins as the officers ride their horses back through the twilight to their quarters. The captain, mistrustful of the morning sky, says: "Well, Weller, what do you think, if only 'Anna' doesn't rain on us."

The horses trot off.

After a while Weller says, "X plus four hours. Maybe we'll get back on the horse and ride to Paris."

The horses trot off.

Reisiger, bent over in such pain that his chin hits the saddle button once, says half aloud more to himself or to the horse, "That damned X!" And now even more quietly, completely inwardly: "How well that's done—and if you can fix every hour in life by the clock—there's bound to be an hour X—the great unknown—where you croak."

Trot. Reisiger can no longer hold on. Behind him rides his lad. He sees the lieutenant slip over the horse's neck on the left. He catches him and puts him on the ground. The horse stops. "Herr Feldunterarzt!"

But they are already far ahead and can't hear.

What's left? Reisiger lets himself be lifted back onto the horse and rides off.

The sun is shining. The quarters are quite friendly. The scribes and draughtsmen are again sitting at the long table downstairs in the study.

When Reisiger enters, Dr. Winkel is already in the room and goes to meet him: "Listen, Herr Reisiger, this is sheer madness. You look like cheese. If you don't want to listen, I'll have to report you to the captain."

Reisiger sits down at the table: "The captain is already asleep. Besides, dear doctor, we have other things to worry about than you senior medical officers." And as Weller enters: "Herr Weller, I think we'll write the final orders now, written only by officers, leaving out 'Anna's' date and the 'X'."

That's what happens.

In the afternoon of that day at about 3:45 p.m., the regimental ordnance officer rudely and unconcernedly pulls open the door. The captain makes an ungracious face. However, he does not get the chance to express his disfavor. The ordnance officer, excited, with a broad smile, says, "Regimental orders: Anna July 15, X equals two a.m."

It is difficult to hold the pencil. Weller and Reisiger draw in the written orders: "Anna equal to July 15, X equal to two o'clock in the morning."

It is barely dusk outside when the messengers gallop to the batteries' ready positions.

A little later, Captain Brett, Lieutenant Weller, Lieutenant Reisiger, and Field Surgeon Winkel ride to the position. In the quarters, instructions are given to pack all belongings immediately and to keep the baggage wagon tensed and on standby throughout the night. All luggage is put on the wagon.

The officers have nothing with them but maps and gas masks.

15

July 15, 1918, 1:30 a.m. What is happening at the front is a monstrous costume party of war. If you followed Captain Brett and his staff, if you joined them when they dismounted and sent their horses with the lads back to the rear, if you now climb with them through the night up the hill in front of you, laboriously, because there is a thick growth of brambles everywhere, if you are now at the top, on the back of the height, you will not understand it; you cannot understand it.

The most monstrous costume party of the war. It's teeming with people. There are more men to the right and to the left and in front of you and behind you than you can count. There's a buzz of voices, whispered half-loudly, but loud enough to keep hitting your ear like the rising and falling of waves.

Look at the men. Are they soldiers? No uniforms, no caps, no steel helmets. Certainly, high boots over field-gray pants. But bare arms, bare necks, bare chests. Are they soldiers?

Strolling around, whistling quietly to themselves, joking, and making jokes.

Yes, it looks as if they are all children, jumping back and forth, playing hide-and-seek, and catching each other, darting back and forth between the scenes of this costume party.

You have to say: the party is well decorated. The festival director spent everything.

As if he wanted to show how much he has.

Walk across the ridge, count your steps, measure the distance: one hundred, two hundred, three hundred steps. Here you must stop, here is a carefully drawn wire fence: the end of the festival hall. And now, just for fun, walk between the jumping children, these three hundred steps back again. And begin, step by step, to count the backdrops, little arbors, little séparés—and a cannon in each one.

One step, two steps: the first row with one meter of space in between, as far as you can see, to the right and left. And, when it gets brighter, further, much further than you can see.

New steps, one, two, three, four, five: the next row. Well done, carefully considered. The second row is, as they used to say in gymnastics lessons, a gap.

New steps. A third row, a fourth row, a fifth row. And if you get tired, and if you could ask the head of the party. Believe me, he'll laugh and say. "But that's nothing yet." And he will spread out his arms, proudly and ostentatiously say, "We have much, much, much more." And he'll raise his thumb and calculate for you: "Please, how big is your living room?"

You think about it and say, "At a guess, six by eight meters." Then he laughs even harder. "Look, in a room the size of your living room there are, let's say, three to four guns here where we both are now . . .

That is." He makes a somewhat tearful face. "But small ones, we have big ones too, of course." And he warms up. "Oh, what do you think, we even have really big ones. So big that one of them would be too big to fit in your living room, which I certainly don't want to disparage; in short, we're a department store." And then he wants to make a joke and pats you on the shoulder like an all-too-confident rayon boss, and says, "There isn't a gun, whatever you think, from 7.5 to 38 centimeters in diameter, that wouldn't be within reach up here this morning." And then he pulls his head back between his shoulders, a little tearfully again, and continues, "However, we can't let you have even one at the moment. We need everything ourselves." This startles him, he checks his wristwatch and mumbles nervously. "You'll have to excuse me . . . it's already ten minutes to two."

He is already gone.

The swarming people in shirtsleeves have now crept up to the séparés.

Always a gun with the crouching, whispering figures. Next to it, higher than a person, carefully stacked, a mountain of ammunition, guns, ammunition, guns, ammunition—no telling how much.

Costume party? Costume party? Costume party?

Somewhere on the fairground, a mast stands erect. A few wires swing from its top. Two people are sitting on the ground next to it, telephone receivers around their ears, motionless.

You step up to them.

Suddenly, one of them jerks his head and says to the other. "Now I hear the buzzer, long, long, short, long, long."

The other says, "That's 1:55 sharp."

They compare the clocks. They splash apart. One to the right, the other to the left. You hear, passed through all the rows of guns. "Precise time 1:55 and ten seconds."

Costume party?

But the children playing have gone quiet.

The whispering has stopped.

Now and then, a voice half-stitches: 1:57. After a while: 1:58.

Costume party? This churning out of numbers is almost maddening. Like a shred through the air: "Nine fifty . . ."

Then. Then there is no more night, no more height, no more forest, no more guns, no more people. Then the whole thing is dissolved into the most terrible roar that has ever broken out of the earth. All you see is a single sheaf of fire, hundreds of meters long and hundreds of meters wide.

This is Operation Anna, on July 15, 1918, two o'clock in the morning.

16

As we are now on the verge of concluding the current major battles in the West, the course of which has brought us a significant success and great victory, we no longer have to fear any surprises from Foch, especially as we have already wrested the initiative from the hands of the Entente through our success in March and paralyzed General Foch's operational army.

The most important success, however, is that we have destroyed the enemy's plans for 1918, and the enemy no longer has the opportunity to seize the initiative.

(Press conference, Berlin, June 8, 1918 / Kurt Mühsam, "How we were lied to," p. 113)

17

Brett and Reisiger are sitting in a hole in front of the first line of guns of III/253. They climbed in ten minutes before two. Since two, the hole has been continuously lit by a reddish fire. The muzzle flashes of the first row of guns are no more than five meters behind them.

They both know: There's no point in being here. This gigantic machine, which is now in motion, can no longer be corrected, can no longer be stopped by any power in the world.

The crashing is so loud that you can't hear a word, even if you shout into each other's ears within centimeters. So, they both sit silently, looking in front of them. They have lists with them and occasionally compare these lists with the clock. Now, they know, the Ninth Battery is firing on the French Battery 217; that will take another fifteen minutes, then the French battery will be a pile of dirt and blood. Now, at the same time, the Eighth is drumming on the French machine gun emplacement "Leopold," only five minutes, that's enough, and it can continue firing, three divisions to the left, four minutes on the "Senta" sapper, then further, another two divisions to the left, on the crest of the large dugout "Emil," which the aerial photographs showed so clearly. Looking at the clock, aha, now the fire of all three batteries is united on the enemy heavy battery "11 Ludwig." According to the list, this takes twelve minutes. Only gas will do here. Hopefully, the battery commanders won't forget to count out the Grünkreuz ammunition beforehand.

Reisiger is still writhing in dysentery pain. Every now and then, he drops the list and lies down. The captain takes no notice. So, he straightens up again and continues his pursuit.

The firing of the batteries is the best way to forget yourself completely. And if you already have a fever, this flashing and this roaring make your blood even hotter. It's a condition, really: fever multiplied. That you become so light, you almost seem to float until the cursed pain becomes all too sharp and presses you to the earth again and again.

Compare the list. Check the clock: Now this is happening, now that is happening, always precisely fixed on the enemy. No carrion will escape. It is completely unthinkable that a single creature can remain alive over there.

The batteries have only been firing for an hour. But, without exaggeration, even now, no enemy can live.

Reisiger looks at Brett. He is slapping his knee, laughing. Reisiger knows what he's thinking: a fabulous war. The infantry afterward is to be envied. They can safely go for a walk as long as the sky is blue. I'm sure they won't meet a single person again.

So go on, three more hours.

Three more hours. It's a painful realization: three times sixty minutes is one hundred and eighty minutes. One hundred and eighty minutes of this horrible crashing and roaring and raving. It's not just your ears that hurt. You gradually become more differentiated. You can already imagine that the air pressure of every single shot is beating against your chest. One hundred and eighty minutes in which—*yes, how many shots are fired per minute?* That cannot be calculated. It cannot be estimated. No matter how carefully you listen, you can't really tell the difference. It's not a gun firing a hundred or a thousand times, but the endless wall roars incessantly, incessantly.

When does it fall silent?

The pain in the ears gets worse. You put absorbent cotton in. But you quickly tear it out again because you have the feeling that it is being chased straight into your brain. And pain in your chest.

And now, as if tiny bullets were constantly hitting your whole body, pain all over your skin. Pinpricks, a burning, tingling sensation. It can't be explained as if the air had been shattered and was trickling down on you.

And Reisiger had repeated and repeated attacks of dysentery, this terrible burning in his body. He hadn't eaten anything for two days and wasn't allowed to drink anything.

An insane thirst sets in.

It's all the worse because you know it's hopeless to get even a drop of water. You can't move at all. There is no chance of getting out of the hole for the next few hours. Even if you lay on your belly on the flat ground, you would simply be burned to death by the muzzle flashes of the nearest guns.

So what consolation is there? The watch and the list. And keep track of what this battery is doing now, and where the gun is firing. And if you want, you can imagine: *How do the crews feel? How do the officers feel?*

And then, you think to yourself, *Is the enemy actually firing?* You can't tell. The thought becomes all the more horrible: *If he shoots, Who is still alive?* Because there is no cover. No one expected that.

Reisiger looks at the captain. He has finally lit a cigar and is crunching it between his teeth. A good sign. So, he seems to be satisfied, feels good. Reisiger envies him: How much more robust he is, that he can smoke now.

Five o'clock in the morning. Lord God, another hour.

Reisiger is beginning to find the tension so unbearable that he has to make an effort not to suddenly jump up and start running. No matter where, no matter what for.

It's better for those who are now operating the guns. And even if the enemy responds, at least they have something to do. This idle sitting around is horrible.

Reisiger looks at his watch, looks at the list. He knows that the fire of all the batteries is gradually converging on the entire front. Only the very heavy calibers continue to pound the enemy's rear area. Everything else joins together to form the fire roll.

The famous fire roll. This curtain of fire and splinters is now erected shortly in front of the first German trench. And then, at 5:28 minutes, begins its march, 5:28 minutes ten meters in front of the German position; 5:32 minutes twenty meters. And then it rolls toward the enemy. Behind it, upright, rifle at the ready, our infantry. Fire roll, rolling forward. Roll forward. There won't be much life left. But perhaps, somewhere, by chance, a miserable person has saved himself. Roll forward—and will now tear him to pieces. Rolls on. Behind him, slowly, almost leisurely, the German infantry. Rolls forward, stops for another ten minutes on the first enemy trench, jumps thirty meters further so that the German infantry can now, in the first enemy trench between the dead enemies, catch their breath a little. Roll on. Now more generously thirty meters, and again, and twenty. Second enemy position. Hold on for ten minutes until the German infantry has approached. Onward, onward over the third position, over the fourth. The Germans are always behind. It must indeed be a walk in the park. No enemy can live. 5:55 a.m. Already far into the rear area, where there are certainly no more trenches. It's 5:57 a.m. Now the fire roll is already jumping 100 meters, will certainly roll through the villages behind the front, will catch the staffs, the baggage, the columns. 5:58. 5:59.

The captain jumps up. Reisiger jumps up. It's six o'clock in the morning. The whole front is silent. There is absolute dead silence.

Reisiger and the captain take a few steps to the first line of their batteries. The barrels are smoking. The mountains of empty cartridges are smoking. Exhausted, the men and officers lie on the ground, hands and faces black, pants and shirts burnt.

Dead silence.

The captain thinks it necessary to say something. But he is as agitated as Reisiger has ever seen him. He can only find one word: "Excellent."

They go to the hole where Weller and Winkel are lying. "Well, gentlemen, fine thing, isn't it?"

They see Winkel in a corner, his head in his hands. He is staring at the ground.

The captain: "Doctor, what's wrong?"

Winkel tries to stand up, slumps down again, and points to the opposite corner. "Lieutenant Weller is dead," he explains, stammering, that they had covered their hole with a canvas so as not to be constantly bothered by muzzle flashes.

A gunner, who was probably looking for ammunition, had not seen the hole and had fallen into it. In the excitement of the night, Weller apparently mistook it for a shot from the enemy. "I can't understand it," says Winkel, "Lieutenant Weller screamed horribly . . . The shock probably killed him . . . . Heart attack."

18

The officers of Battery III/253 gather around Captain Brett. They are standing in their shirtsleeves, because they have also fired or carried ammunition.

Yes, and now? What's going on? There is no talking. Every now and then someone steps to the side of the hole and takes a look at the canvas under which Weller is lying. At one point, someone says, "That's probably the only dead person. Really bad luck. So close to the end." Everyone looks questioningly at the captain.

What can he know? He knows nothing. He doesn't answer either, just looks at his watch, and stomps back and forth.

Ten minutes pass. Something must be happening now. At least one officer of the High Staff behind the front, who were supposedly watching over the attack during the night, could appear here and say something about the result, about the progress.

Or perhaps the Protzen will come soon. We know that they are supposed to arrive automatically one hour after the end of the artillery fire. So that they can limber the guns—and then march forward, march!

Don't the radio operators know anything? They must be in wireless contact with the army group.

An officer is sent to where the mast with the antenna is. He comes back: nothing known. Nothing can be received at all. Apparently, the enemy is transmitting interference waves.

Hmm. Isn't there any infantry coming? It's clear that the reserves must now gradually move forward. Because everything that was in front will now have stormed many kilometers away, forward, and connections to the rear must be established somewhere.

Officers are sent out. Come back: Yes, you can see larger columns. That means they're in the open, so they probably don't have orders to advance yet.

6:30 a.m. The captain is getting tired of waiting. As he looks to his right and left, there are other staffs everywhere, groups of officers. Indecisive, like him and his masters.

He finally turns hard on his heels. "Herr Leutnant Reisiger, please take a sergeant and go to the front. Try to clear things up. It's ridiculous that we're being left sitting here. I want to know approximately where the first line of our infantry is." He turns to the other officers. "After all, they're calling for fire up ahead, and you're shooting into your own men. It's too silly for me. I'm not going to wait any longer. Do you understand, Herr Leutnant?"

Reisiger fetches Sergeant Boll, whom he knows well from Russia. Not older than him, dashing, calm. "Boll, take the gas mask, any rifle, go." Greets, clicks his heels, walks forward.

"I want you to send the sergeant back if you think you have to stay in front for more than an hour," the captain calls after him.

They turn again. "At your command." Reisiger and Boll disappear.

19

They go downhill for a few hundred meters.

Reisiger feels a little better. Now that the sun is rising, where you can see the green grass, quite high, with lots of poppies, he even feels quite comfortable. He starts moving faster and eventually starts running. He enjoys being able to move again. Boll is next to him. The gas mask is still in the canister. The rifle hangs on the belt hook, rattling and drumming a merry beat.

"It's fun, isn't it, Boll?"

Boll is a little out of breath. "Yes, Herr Leutnant, definitely better than in Russia."

Reisiger runs on: "Look at all the poppies. This patch of earth seems to have no idea what was going on above it this morning ... I was pretty heavy, wasn't it?"

"At your command, Herr Leutnant," Boll says this proudly and with a little grin.

Still downhill. Then the terrain gently rises again. A birch forest comes closer. Still young trees, maybe three meters high. The sun shines against the white trunks. It looks beautiful.

Suddenly, Boll stops and holds Reisiger by the arm.

"What is it, Boll?"

"Excuse me, Herr Leutnant, I don't think the forest is completely clean. The leaves aren't green, they're purple."

Purple leaves? Reisiger stops. But that looks very strange. Indeed, purple leaves. At least of a color that a birch forest certainly doesn't have. Without any trace of green at all. "So, what, Boll? You're thinking of gas?"

Boll nods. "At your command, Herr Leutnant, gas."

They slowly walk closer. *Gas?* Reisiger thinks. *No battery of ours could have fired here.* "Did you hear any enemy batteries this morning, Boll? I mean, the forest could only have been gassed by the enemy?"

A few more steps closer. It looks horrible. Now you can see that even the white trunks are splattered with a purple-red greasy liquid. And you can see small, shallow blasting holes. Fresh soil. Yes, so the enemy must have fired without being heard.

Map out. We have to go through it. The only question is, where is the shortest way to get back to the open field? To the former German trench?

"Boll, that's not so bad. The whole birch grove is no more than a kilometer deep. We can do that in ten minutes." Gas masks on and into it.

That's what happens. It's difficult to decide how best to behave. One thing is clear: Get through as quickly as possible, preferably at a gallop. *But how can you cope with the gas mask on?*

Another thing is clear: For heaven's sake, don't bump into any trees. Don't touch a leaf. Put your hands in your trouser pockets. Make yourself as tight and small as possible. Then you'll be fine.

You are already between the birch trees, seeing nothing but the gassed forest in front and behind and on all sides. With the purple leaves, the red splashes on the white trunks.

Perhaps it would be better to walk very slowly. Or even better, to crawl. Because the whole thing is terribly eerie, it's as if everything is under a heavy bell jar full of oil.

Reisiger squints upward. He has to stop. *This is*, he thinks, *a desecrated forest*. These are trees, birches ... three or five years old. They have nothing, nothing, nothing to do with the war. Who don't want to decide to side with the Germans or the French. Who do not agitate, do not murder. Who just stand there and get their leaves and blossom every spring and lose their leaves in the fall, and freeze patiently until the next spring. Without haste. Animated by nothing more than, perhaps, the urge to have sun.

And now? Now, the biggest beasts on earth, the humans, have attacked these miserable birch trees.

A whim has taken hold of this forest. And it dies as wordlessly and as devotedly as none of its murderers. Admittedly, there is a bit of wind, and so the trees are still shaking their heads a little. But all the branches have already stretched and bent. And the leaves trickle and trickle. And it won't be another twenty-four hours before there will be bare poles here. And all because men wanted it that way.

It's childish to dream here. Boll is ahead. Many steps. He turns around and stops. He says something. You can tell by the way his gas mask inflates. Reisiger can't understand anything yet. But he nods his head and follows. The two trunks of the masks clatter against each other. "What do you say, Boll?"

"That's a mess." Boll squints against the trees through the large glasses of the gas mask.

"Yes, Boll, but first, just get out of here."

They walk on, with Boll in front of Reisiger, very cautiously. Finally, up ahead, the forest thins out. You can see brownish sand, and above it wisps of blue sky.

Reisiger on Boll's heels. "Quick, Boll, we'll soon be out of this mess."

Boll jumps forward.

Suddenly, Reisiger notices that a branch of a birch tree is protruding far into the track. *Can't Boll see that?*

"Boll!"

The branch tears open the sergeant's mask behind the eyeglass over his entire right cheek. The rubber skin opens up like a large wound. Boll raises his hand and covers the wound. He rushes on. Reisiger behind him. Three more steps. There's the open field.

"Boll!"

Boll opens his arms wide. Does he want to welcome freedom after this prison?

But he falls to his knees, very slowly. Gracefully.

Reisiger jumps up to him. "Boll, what's wrong with you?"

The whites of Boll's eyes suddenly turn deep red. There are lots of white blisters on his mouth.

"Boll!"

Boll is on his knees. His upper body slowly falls further and further back. He bends, and his head finally behind his boot heels.

"Boll!" Reisiger shakes him. If only he wanted to say one more word!

There is no way of forcing this one word out of him.

There must be someone around here. I know for a fact that there are hundreds of thousands of infantrymen lying here. And surely medics. With oxygen blowers. I'm sure there are excellent treatments for gas patients. *But should I leave Boll here alone for so long? Or do I have to try again?*

"Boll, should I get a doctor?" Reisiger is still kneeling beside him. Still sees the red in his eyes. *But why so complicated?* he thinks. *Feel his pulse.* "Boll, let me take your pulse."

He takes Boll's hand. But he drops it again much more quickly than he took it. It's no longer a hand. It's lifeless.

"Then I'll just have to go on alone. Too bad . . . Boll," Reisiger says this out loud. Then he trots forward.

It's an open area, perhaps once a farmer's field. Milling sand, and a very good, quite wide path here.

Where is the infantry? And is it even possible to walk upright here?

Map out, compare the area. Right, there's a new clearing on the right. Birches again, very small ones. Reisiger, if you go through there, you will inevitably come across the second line. And there are trenches. And then it goes on.

God, this gas mask! It's soaking wet, breathing water is always collecting on the right and left. And now it washes against your chin as you run and sometimes spills into your mouth. Yikes!

And then the glasses. You can't really see anything anymore. It's like the whole landscape is covered in fog.

But if you don't have the mask . . . What a bad luck with Boll.

Reisiger pauses as he walks. *Something is very strange*, he thinks, *you can't hear a single shot.*

This thought is usually quite comforting. But now, here, it frightens him.

As an old warrior, you only tend to fall on your stomach when it shoots. Reisiger finds himself falling down because it's not firing.

He lies flat as a board in the middle of an open field. The sun is shining. The trunk of the gas mask is almost digging into the sand. He is very aware of it all, and for a moment, it seems highly ridiculous. *Man in the countryside*, he thinks mockingly. *The war is a funny invention. My parents should see me like this, here on French soil. In the morning sun, all fours stretched out, with my trunk in the dirt.*

It would be nice to lie here, at least sleep. Nobody's shooting after all. Who knows where our infantry is. Maybe they've already got it good, in the provision offices, in the enemy positions. The English have whisky. And by the way, it's very appealing to be able to see for yourself once again what butter is.

No, no, that's not how it works. Dear Reisiger, you're not sunbathing here. Yes, just squint. There's another little wood up ahead. You have to go through it. There'll be an infantry post up there that can give you the approximate route to Paris. So, let's go!

The few minutes were relaxing. *You must talk to yourself from time to time*, Reisiger thinks. *It increases your so-called fighting spirit.*

He stands up, carefully pats the sand off his skirt and trousers, and walks on.

Not a shot is fired.

Yes, and what is that? How he reaches the birch grove, the small one: no purple leaves, no splattered trunks.

This is a real, normal forest, just like the one you have at home. It's a miracle. So now through it. We will reach the German infantry in five minutes.

Reisiger takes off his gas mask. He does it like a little private ceremony. He makes a slight bow toward the birch trees, pushes the steel helmet off his head from behind, and slowly removes the cursed rubber hood from his face with the movements of a priest in church.

He wants to throw it off first. No, no, it's already excellent. He slowly and solemnly places it in the steel helmet. *Now it's my turn.* He puts his hands on his hips and straddles his legs, catching his breath—a proper, decent, clear breath. A deep gulp. A deep exhalation. *Thank God I'm still alive!*

Then the gas mask is dried with a dirty handkerchief. Hanging it on the chest as a precaution. Steel helmet on. Gallop through the forest.

Right through: The rampart of the rearmost German infantry position is piled high.

"Guard!"

Nobody answers.

But of course, no one can report. That would be pathetic. The men have moved out, who knows how far forward.

Reisiger jumps into the trench. Curious. I'm the first artilleryman to make contact with the good infantry. Well, they'll be pleased. I wonder what they'll say?

So, here we go around the corner. Through the trench and now trot forward! It has to be said: a well-developed position.

You almost have to sing. Everything is so clean here, with good grates on the ground. You can rumble on it. That's fun.

Forward. Aha, already a cross trench. Anyone sitting here?

"Guard!"

Reisiger looks to the right: nobody. Then to the left. He walks a little way in. Here is the first bunker.

"Hello!"

Nobody. So back into the trench. Gosh, they seem to have run well. Yes, yes, Operation Anna. That's how the artillery works.

And trot on.

If only it wasn't so warm. But after all, I've been trotting through these narrows for a good twenty minutes now. It's gradually getting boring. Nevertheless, he trots on.

Forward, forward! And finally, after crossing a third trench, he sees a fourth from afar. A high black wall. The breastwork is particularly high. Aha, that must have been the former first trench. If only I could find a few guards here. They'll be able to tell me how far the main troops have advanced.

Of course, something's moving. "Hello!" Gosh, it's an officer. An older officer. He won't be very pleased if a young lieutenant shouts "hello" at him. So, hurry on and get to him quickly.

Reisiger almost stumbles as his feet clatter over the wooden grating.

He stands two steps in front of an officer with a gray beard and gray hair. Well, without a helmet?

He moves even closer and salutes. Right, a major, the epaulets are supposed to be covered, but one of the covers has slipped a little, and you can see the thick caterpillars.

"Leutnant Reisiger, Ordonnanz Offizier III. Division Field 253 obediently reports for duty."

The major doesn't move.

His hand still on his helmet. "If you allow, Herr Major, Ordonnanz Offizier Reisiger . . . from the field artillery."

The major takes a step toward him, raises his hand as if to grab him by the collar, then drops it. Grins broadly. Then a horrible, screeching laugh comes out of his mouth.

Yes, what on earth is going on? Reisiger is at a loss. *I wonder if he doesn't understand me.* He puts his hand on his helmet, still according to regulations, even if his voice is a little shaky: "My commanding officer is asking Herr Major for an approximate indication of where our infantry is now."

The major raises his hand again. But now it is not toward Reisiger. Instead, his arm bends, and the back of his hand passes abruptly over his eyes. And then it sounds like a laugh. And then Reisiger realizes that this laughter is a terrible crying fit. Tears pour out from under the back of his hand.

What on earth are you supposed to say? Has he gone mad? How can you make yourself understood? Oh nonsense, military forms.

Reisiger moves close to him and wants to put his hand on his shoulder. He doesn't dare but says, "Can I help Herr Major in any way?"

The major's hand falls away from his face, dropping downward. Big gray eyes look at Reisiger. Keep looking at him. Still looking at him. Finally, the head moves. Then the shoulders jerk up. Then the head turns to the side in the direction of the trench. Then the shoulders turn back. Then the torso turns with difficulty. Then the left arm attempts a helpless movement, also pointing in the direction of the trench.

Reisiger steps past something, and looks in that direction. "Herr Major!"

Not another word.

An infantryman with a white face is lying on the breastwork, knees on the assault ladder. One more next to him. A third. A fourth. A fifth. Ten. One hundred. As far as you can see; one man next to the other. Always the head fairly high, the hand on the rifle. Always the left knee on the last step of the wobbly little storm ladder. And always a small hole under the steel helmet, between the eyes, or in the cheek, or next to the ear, or in the neck.

There's no point talking to the poor major. Reisiger pays him no further attention. He goes into the trench. His head is about the same height as those strangely petrified knees that lie on the last step of the storm ladder.

He walks on. Abandoned bunkers. Tins of canned food. Hand grenades are neatly arranged in large piles everywhere.

He walks with his eyes closed. He no longer looks because he feels it—you can go as far as you want. There's always a knee hooked into a storm ladder at your left shoulder. There is always a foot in an uncleaned, dirty boot at the height of your left hand.

But there must be some living creature here in the trench!

He goes on. He wants to run. He has no courage to do so. He walks slowly, step by step, his eyes to the ground, on his left shoulder.... It is impossible to imagine. Sometimes he whispers into the entrance of a bunker: "Infantry." The whisper also echoes. But no answer.

He is happy to finally see a sergeant hanging across the trench, head down, knees hooked into the storm ladder. A different sight at last. He's dead too. But at least he has a sense of humanity.

Next.

Here comes a sapphire. Is someone alive? He hesitates to go in. But then he does. It's only a few steps. A machine gun. The sergeant, head down: dead. Two men, also ... No, no, one of them is alive! Of course he's alive! He's lying on his side, without his steel helmet. He has his free hand tucked into his torn skirt. He is breathing.

Reisiger slowly approaches. "Well, comrade?"

Silence. It lasts a while. Then his eyes widen. That's his answer.

"What's going on, comrade?"

Truly, he has understood. He half sits up. Reisiger grabs his arm around his neck and holds him.

"You can believe me ... it's over now .... Whatever came out of the trench lies a hundred meters in front of us ... dead, of course .... And the rest of us ... well, there you see ... Can't you take me with you?"

Reisiger puts him down again, strokes his head. "Is the enemy still ahead of us?"

Answer: "He must be. I've never seen ... so much machine gun fire ... as this morning ... take me with you."

Enterprise Anna?

"Comrade, you have to take me with you."

Take me with you?

A machine gun starts firing...That's the enemy. So that's what we spent five hours drumming for? What do you mean, take me with you? The best thing to do now is to stand here, legs apart, over the machine gun...no, no, the captain is waiting for news.

"Comrade, I can't take you with me. Shall I send you a medic? Besides, your major is still here."

"Take me with you!"

All right, then. But how? "Can you climb on my back?" The infantryman straightens up and looks at the dead sergeant lying across his feet. He smiles.

Oh yes... Reisiger pushes the dead sergeant aside. He stumbles into the trench. "There, now this will work."

The infantryman sits, spreads his legs. Reisiger pushes his back against him.

The machine gun barks again. Reisiger ducks down because it seems as if the bullets are flying around his ears. Then he hears the infantryman on his back let out a muffled scream and throw his head far forward. A jet of blood as thick as a finger shoots past Reisiger's cheek. The infantryman falls heavily to the ground. That was the machine gun.

Now Reisiger forgets everything. It doesn't matter at all whether anyone is still alive. Even the dead are of no interest.

He races off. The major kneels there and cries and cries. That's just as uninteresting. And Reisiger turns his back on him. And he hears the screams behind him. But that's completely bogus. He roars like a bull. Who cares. And the one machine gun that has been barking until now suddenly seems to have turned into a hundred. That doesn't matter either. All the better to run.

And through the trench and through the birch grove and over the sand and the steel helmet off and the gas mask on and past Boll and sneaking like an Indian through the desecrated forest and with the sides of the gas mask puffed out across the field and up the hill and gas mask off and jumping to the captain, who is sitting comfortably on shell baskets in the circle of officers, puffing his cigarette. "Herr Hauptmann, everything is dead up there. The attack failed." And fell over. And forgot about the war, the world, and life for ten minutes.

Soaked in ammonia. Standing over him, Winkel, who makes a motherly face. "Good morning, Herr Reisiger, you're waking up at just the right time."

There's already the command: "Put the batteries in march." There's already the lad with the horse. In broad daylight, the command is already blaring from the voices of many officers: "Battery march!"

It goes backward.

You can already hear the furious barking of machine guns behind you on all sides. It's the enemy.

Smoke and fire are already billowing everywhere in front of you. That's the enemy.

20

Our mensura scares[20] are often not understood by the public. But that should not mislead us. Those of us who have been corps students, like me, know better.

(Wilhelm II, May 7, 1891)

21

July 16   *Western Theater of War*

Army Group German Crown Prince:

Southwest and east of Reims, we penetrated parts of the French positions yesterday morning. Observation and survey troops played a special part in the preparations for the artillery battle. Artillery, mine launchers, and gas launchers, together with armored cars and flamethrowers, opened the way for the infantry into the enemy.

---

[20] Mensura. Academic fencing, also known as Mensur, is a regulated duel between students at university fraternities, utilizing weapons such as épées or rapiers, often resulting in severe facial scars. Still practiced today at German and Austrian university fraternities.

22

Already during the fire, which took place from two to six in the morning on July 15, rumors were circulating among the artillery: The plan of the attack had been betrayed by an infantry deputy officer who had defected.

23

This time, the French were not taken by surprise. Foch had known what was coming since July 1, 15 days before the offensive began. Defectors and prisoners gave him all the details. He hurriedly brought up reserves and ordered the armies to structure themselves more according to depth than had previously been the case. The first position was vacated . . . On July 15, as in previous battles, the German artillery and minesweepers smashed the front trenches, but this time they were almost empty.

("The Great War," brief account based on the official records of the Reich Archives by Erich Otto Volkmann, Berlin 1922)

Chapter VI

1

The troops of Boehn's army that had advanced furthest to the Marne have been withdrawn about ten to twelve kilometers to new positions between the Marne and Vesle. The army report will only be published when the enemy has noticed the rearward movement. Nothing must be reported before then. If the Entente, as is likely, is inflating this movement into a great success, there is an urgent need to use the press to influence a correct assessment and to reassure the public. We had intended to break the deadlock created by our advance to the Marne by a double thrust at Reims and to link up with our Champagne front. We were unable to achieve this intention. We had to halt the offensive, which lacked the element of surprise, in order to avoid losses. The enemy's attempt to break through between Reims and Soissons was defeated, so that both parties did not achieve their operational intentions. In considering the situation, however, the German plans should not be mentioned, but it should be emphasized and highlighted particularly strongly on the enemy side. A completely new situation has developed since July 19 . . . We have the advantage of a shorter front, troop savings, and improved rearward connections. In any case, the press must warn against a false assessment of the mere possession of terrain. Our situation has improved considerably for the continuation of the battle . . . The general plan of operations, as we laid it down in the winter, has not been changed by the present events; it is being carried out according to plan.

(Mühsam, How we were lied to, p. 116, press conference July 29, 1918)

2

On the morning of July 15, Regiment 253 had received no other orders than to go back. There was no information as to why or what for. This was perhaps the main reason for the uncertainty that made itself felt among officers and enlisted men, and which was everywhere, tormenting, confusing, paralyzing.

The daytime march soon had to be halted. The sky was clear and gave good air visibility. The enemy pushed through to the hinterland with squadrons of planes, and dozens of machines. This made it advisable to disperse the regiment's batteries, which were marching along separate roads, and to conceal the vehicles until dark as well as possible between ruined houses and under woodland.

The heat was sweltering and numbed the people. There was already fatigue from the early hours of the morning. Now this pressure increased and became a will-less pessimism among the men and officers.

Everyone had realized at the start that the "Anna" operation was more than just normal positional fighting and the usual artillery skirmishes. So why had they gone back? What had gone wrong? The machine had worked just like on the training ground. Every shot had been carried out according to the last detail of every plan.

So now they lay next to the guns, half-dawning, and listened up ahead: Why did we go back?

And the guards among them: The whole thing has gone wrong. Watch out, the enemy is coming after us.

And they listened ahead again: There is little firing. Are those still German batteries? Because when we withdrew, others were still standing.

And then lay down: It doesn't matter at all. We do what we're ordered to do. The main thing is that we finally get something sensible to eat again.

The officers were beaten. They said nothing. But hadn't they all been in the church where the major spoke, the man with the hard voice? "Our operation promises success for everyone." And the first success was that the 253rd Regiment had been forced to charge backward, wasn't it?

And the men of the III Battalion?

Lieutenant Weller is buried, just now, along the chaussée. There is still Captain Brett, Reisiger, and the field surgeon Winkel. And the captain, who has so far been speechless, is now speaking. Strangely obstinate and somehow offended. "Herr Leutnant Reisiger, you've known for a long time that I have gallstones."

Gallstones? No idea. Looking at Winkel. *No idea.*

The captain continues, "And it got so bad during the night up there in that shithole that unfortunately I have to start the cure today, which has been arranged with the regiment for a long time."

Reisiger and Winkel look at each other again without a word.

And the captain pulls out a pocket mirror, examines his fat, rosy face in it, and says, now quite offended: "I'm already completely yellow."

Reisiger wants to reply, but doesn't have the courage.

This obviously annoys the captain. He snarls. "Herr Leutnant Reisiger, please make sure that my lad clears the car with the officers' luggage for me immediately. That means . . . the gentlemen will have to see that their suitcases are stored somewhere else. I'll drive past the regimental staff and sign out; I'm in such pain that I can't ride anymore. And I'll see to it, Reisiger, that you become adjutant and that a substitute comes for me. I will also have Lieutenant Schlumpe of the Ninth appointed as our ordnance officer for the time being. I'll be back in four weeks." He suddenly makes a face as if he is already dying. It looks all the funnier because he is very fat and has sly, amused eyes.

Well, what can you do? Never before in the war has a lieutenant been able to order a captain to do anything. Reisiger probably wouldn't have ordered anything; perhaps the captain is right. Perhaps he is doing the wisest thing one can do. After all, the war is—

"I mean, I'll be back in four weeks if the war hasn't been called off by then." The captain is afraid of any rebuttal. He quickly turns away and sits down against a tree. "So, Reisiger, hurry the wagon, I'll rest for a moment."

After an hour, *Yes, did he actually say adieu or not?* The captain drives off with the lad. The lad, of course, must go as far as Germany to look after the suitcase.

Reisiger and Winkel watch the car. "Don't hold it against me . . . the captain is a swine," says Winkel.

"But, dear doctor, if I were captain . . . I don't know if I wouldn't do the same. Be honest; do you still find any sense in this boundless nonsense?"

Winkel shrugs his shoulders: "Well, at least there is a point: to stop the enemy. To make sure he doesn't get any further."

Reisiger, tired: "Well, you weren't at the front this morning. I know what happened. It's over! I can imagine the High Command calling the whole mess a victory. But I'm afraid this is one of those victories that will break our necks. Believe me, dear doctor, we will kill ourselves to victory."

Winkel chews his fingernails. "All right, victory through death. That's fine with me. But then it's important that the enemy also kills itself to victory."

Reisiger, quickly: "You're preaching utter nonsense. Right, if two parties fight each other and both stay dead on the pitch, then there's peace.... Doctor, I can't help it, I no longer believe in the productivity of this occupation. I doubt more and more that a man's task should be to die, I say. It is becoming increasingly unclear to me whether the meaning of life is really death, or whether we humans are committing the most egregious sin against life when we die so senselessly."

"But Herr Reisiger, who is talking about senseless death? We're talking about dying for our Fatherland."

And Reisiger: "I'm talking about living for the Fatherland. As I said, I don't want to decide anything ... because if I did, I would have sat in the car with the captain ... but I don't know how long I can continue to believe, in private, in dying with honor on the battlefield."

Winkel fell asleep over Reisiger's last sentence. He has opened his mouth slightly, his face is pale and haggard and dead.

Also an answer.

3

July 19, 1918 *Western Theater of War*

Army Group German Crown Prince:

The battle has flared up again between the Aisne and the Marne. The French have begun their long-awaited counter-offensive there. By using the strongest squadrons of armored vehicles, he initially succeeded in breaking into our forward infantry and artillery line in some places and pushing our line back.

July 20　　　　　　　　　　　　　　　　　　　*Western Theater of War*

Army Group German Crown Prince:
Between Aisne and Marne the battle continues . . . During the night we withdrew our troops south of the Marne to the northern bank of the river unnoticed by the enemy.

July 21
Yesterday's day of battle, in terms of the performance of the command and troops and its victorious outcome, was on a par with the great battlefield successes achieved earlier in this terrain . . . During the night, undisturbed by the enemy, we took back the defenses in the area north and northeast of Château-Thierry.

July 22
Yesterday's day of fighting again resulted in a complete success for the German forces.

July 29
Our forward battle area cleared as planned, . . . Our front troops withdrew to their new lines as ordered.

To the German people

The fifth year of war, which rises today, will not spare the German people further privations and hardships. But whatever may come, we know that the hardest is behind us . . . Sacred duty demands that we do all we can to ensure that this precious blood does not flow in vain.
　　God with us!

July 31, 1918　　　　　　　　　　　　　　　　　　　*Wilhelm I. R.*

To the German Army and the German Navy

In the West, the enemy was severely hit by the force of your attack. The field battles won in recent months are among the greatest glories in German history... We are not frightened by American armies or numerical superiority, it is the spirit that brings the decision.

July 31, 1918   *Wilhelm I.R.*

To push back our line . . . we withdrew our troops . . . we laid back the defense . . . Back, back, back: complete success of the German forces! It is the spirit that brings the decision! . . . .

4

No, no, you don't give it much thought when you sit at the telephone by yourself at night and hear the army report go over the wire. No, no, you don't give it much thought when decrees and appeals come in.

But something settles in every night, something crumbles away from every report, from every call.

No, no, you don't think about it. There is no time for that.

That would be thinking too far ahead, and here at the front you only think about the immediate, the daily, the decision and the care for the minute. Every day is broken down into a mosaic of twenty-four hours of sixty minutes each; every minute is the battlefield. Requires action.

At least that's how Reisiger feels.

Captain Brett is gone. The deputy is a captain of the Landwehr.[21] With white hair, wrinkled face, and shaky hands. You can only feel sorry for him; he's so old and so insecure. Really, a miserable fragment of the homeland. It's incomprehensible that there was someone who put the uniform on this old man.

---

[21] Landwehr. Third category of the German Army after the regular army and the reserves. The Landwehr was made up of older soldiers intended primarily for occupation and security duties rather than combat.

And now it's difficult: the adjutant, Lieutenant Reisiger, leads the troop over the commander's head. The captain signs his name under every report and nods to every order. His eyes are always uniformly helpless. He is grateful when he has little to write and little to nod.

With him at the helm, the battalion is now thrown back and forth. Changing positions, once or twice a day. "Herr Hauptmann, the Seventh Battery reports eight casualties this morning."

"Yes, Reisiger will have to find replacements."

"Herr Hauptmann, the Ninth Battery can't get any ammunition."

"Yes, you'll just have to get in touch with our ordnance officer."

Roused from sleep at night. "Herr Hauptmann, a sergeant has just arrived ... the Eighth Battery was overrun by Americans an hour ago. Who wasn't killed has been taken prisoner."

"Yes, Herr Leutnant Reisiger, report this to the regiment. Immediately. We have to call for a new battery as quickly as possible."

After hours of difficulties, Reisiger gets in touch with the regimental staff. The regimental adjutant growls at him: "I can't cut guns out of my ribs either. Besides, we haven't been able to get any material for days. The III. Battalion will stay in position with the other two batteries."

But forty-eight hours are enough. Then it turns out that the other two sections have fared no better. On the same day, the First Battalion loses two batteries and the Second Battalion one.

And the regiment, reduced in head count by about 50 percent, is pulled out of the field.

5

Chief of the General Staff of the Field Army.
Ia No. 9845—to be kept confidential

Secret! Written by an officer!

G. H. QU., August 16, 1918

To the Royal General of the Artillery, Minister of State and War, Herr von Stein, Excellency ...

By decree of the Ministry of War No. 1101 /7. Eighteen, it is announced that for a number of offenses, the previous minimum punishment of fourteen days of strict confinement will be replaced by fourteen days of medium confinement. In addition, the effect of the law is also to be extended to sentences not served before its entry into force. I do not know whether this law has already been issued.

However, after listening to several army leaders, I can only caution against the entry into force of this law and against any mitigation in the use of punishment, as both no longer correspond to the true interests of the field army. The army is increasingly calling for the reintroduction of the punishment of tying up for cowardice or other serious offenses, which, unfortunately, are now quite common.

This wish is all the more justified because, despite all indications, despite the decree issued by the Ministry of War under July 22, 1918, No. 7385/18 C 4, our courts are still inclined to such a lenient application of the law, which does not correspond to the degree of indiscipline that actually exists in many cases. The damage caused by such an incomprehensible administration of justice is irreparable.

Signed v. Hindenburg

6

It is difficult to bring a regiment into a resting position. For where is there rest? In the past, yes, when the front was stable, when the enemy's artillery was built in, when their firing distances could be known or calculated, the peaceful terrain lay kilometers behind the lines.

What is rest today?

There is still a formation in a forest that is at peace. Ammunition wagons or the wounded or the dispatch riders are still coming back: Oh, the enemy is far away, the shots of his artillery are a good half hour away from the forest.

And you can check your watch, and you see it again and again; now the shots are only a quarter of an hour away. And it doesn't take long before the resting place is already the battlefield. Retreating columns are breaking in between the resting batteries. What does rest mean?

And in the past, of course, sometimes an airplane came. Of course, one bomb or two or three could kill and confuse. But you took cover as best you could. You knew: What can a plane do? Drop the bombs, two or three, hit or miss. Then the plane has to turn around and fly home. And it will be quiet again.

But now and today? One plane has become a column of twenty. Densely massed, wings almost next to wings. Two or three bombs have become sixty to a hundred. And when they are dropped and when the swarm turns back, the next one pushes forward, wing after wing. This goes on day and night. Where is rest?

It is rarely possible to really rest for even twenty-four hours.

The III Battalion seems to succeed.

There is a forest. A clearing in it, as big as a marketplace. A pond in it that provides water. An ideal bivouac. The batteries are shielded. The horses stand on lines between the trees, chewing moss and tree bark. Along the edge of the forest, a few hundred meters away, tents are erected for the crews and officers. Well out of sight from the air, but with a view of the clearing and the sun. With access to the lake. So, with the opportunity to cook carefully.

Two more tents away from the batteries: the staff. The captain with Reisiger, Schlumpe and Winkel.

Always the thunder, always the thunder. How well you get used to it. It is far away, sounds like a thunderstorm.

Hours ago, the battalion sent one wagon of each battery further to the rear. They have been told that there is a provisions' office there. So, we quickly receive as many rations as possible. The last few days, the last few weeks, consisted only of greasy potato bread and crunchy tinned sausage. Maybe we'll get some meat for once? Or some cheese? Or fish?

But the crews don't like to wait until the wagons are back. What for do horses die? Wherever the batteries passed a dead animal ... and many roads were lined with them ... the gunners jumped off the guns during the march, cutting off horse meat in large shreds.

This now roasted on small, mysteriously concealed cooking holes. Eaten, half raw. That's good.

The two wagons come back: there's nothing but potato bread and sugar, yellowish, melting stuff. But—roar—a small barrel of schnapps for each battery. Yes, that's an explicit order: Schnapps is to be distributed.

They pounce on the barrels.

Halt! The same thing happens with both batteries: The battery commander says stop. No, no, of course it doesn't work like that. Counting: There are two cups of schnapps per head in the barrel. That's too much. Everyone will get a cup full tonight. And the second tomorrow. Who feeds the *staff*, by the way? They get extra from the Eighth Battery car.

Yes, they get extra. The captain's lad points a cooking utensil through the opening of the tent where the officers are lying in silence, grinning: "Herr Hauptmann, there's arrack today."

"Is that all?"

"Herr Hauptmann, there's still bread. And sugar. If it's all right with Herr Hauptmann, I'll bake some pancakes. I still have some margarine... And if the gentlemen want, some grog afterwards. It's already getting a bit cold..."

As Herr Hauptmann says, "It's done."

Reisiger fills the drinking cups from the hot cooking utensils. "May I take the liberty, Herr Hauptmann..."

Then half aloud, more to himself, "Oh yes, nice is different."

Lieutenant Berger from the Eighth Battery comes running up. Hand on his cap. "Allow me, Herr Hauptmann, to have a word with Lieutenant Reisiger." Reisiger stands up and takes Berger to one side. "What is going on?"

Berger, still out of breath: "Reisiger, the Eighth mutinies."

Mutinies? "'Mutinies'... well, well, Berger, be careful. It seems to me that this word should be used with caution."

"Come along quickly, Herr Reisiger."

They leave without saying anything to the captain.

7

The closer they get to the bivouac site of the Eighth Battery, the louder the noise of many voices and the singing becomes, which had already been heard quietly at the staff tent.

Finally, the bivouac site is visible, and soldiers can be seen staggering excitedly back and forth between the tents. Berger turns to Reisiger and points with his hand. "There are the bastards."

Yes, that does look nasty. "Are you the only officer, Berger?"

Berger, half-important, half-desperate: "Of course. Lieutenant Ziegel was killed this morning. I'm battery commander and observation officer and whatnot . . . but that's no reason . . . . People just don't want to do it anymore."

They are at the bivouac site.

Infantry are on standby in the forest. The infantrymen are startled by the singing. The place is lined with them like a circus. They stand without their rifles and belts, hands in their pockets, staring.

Reisiger and Berger push some of the people aside and step into the arena. Reisiger in front of Berger, in an overloud voice: "What's going on here, have you gone mad?"

That makes no impression.

The master sergeant sits like a bird on a perch on the drawbar of a Protze, next to him the staff sergeant, next to him three other non-commissioned officers. They bob and sing, roar—you can't quite decide which song they are singing. They have glazed eyes. The steel helmet hangs from their necks. A sergeant has lost his helmet and is ruffling his red-haired bristles.

Gunners sit and lie at the feet of the swaying group. Singing along. They take no more notice of Reisiger and Berger than to squint upward with their faces askew and their mouths slightly contorted. That's all.

Only the infantrymen are curious about the outcome of the scene. Just opposite Reisiger stands a deputy officer. As Reisiger looks at him, he takes a firm stance and clicks his heels.

Reisiger turns half-left to Berger. "The whole gang is drunk."

Of course the whole gang is drunk. But what excuse is there for that? Besides, why are they drunk? Berger feels something like an accusation and says to Reisiger, "Yes, it's the sergeant's fault. He just let the barrel of schnapps get out of his hands, and then the whole gang started a fight. And the barrel is empty. That's when the mess happened. And he joined the drinking . . . . I want to talk to him again."

He walks toward the master sergeant. The closer he gets, the louder the roar, the more the drawbar rocks. The chains clank.

"Sergeant!"

No answer.

Berger steps within two paces of him and shouts so that his body trembles: "Sergeant, I'm talking to you. Stand up, for God's sake."

Some of the lying gunners stand up, fall silent, curious to see what is about to happen.

The singing becomes very loud; the notes are drawn out twice as long. The sergeant does not react. He just puts one arm around his neighbor's neck. The others hunker down. So, it's a wall. How can you break through it?

Reisiger wants to jump next to Berger in one leap to put an end to the situation in some way. Berger grabs the sergeant by the collar and pulls him against him.

Aha, the singing stops. Silenced with a jerk. The sergeant stands. He is a good head taller than Berger. It looks helpless, the way the young officer has this much older man by the neck with his arm outstretched.

The sergeant makes a small defensive movement backward. The steel helmet falls off his head, hits the drawbar, and the ground.

The sergeant slowly takes both of his arms, crosses them over his chest, and pushes them upward, thus forcing himself out of Berger's grip.

Still dead silence.

Reisiger is overcome by animal rage. He is now standing next to Berger, shouting "Sergeant! I'm telling you ..."

A red-haired non-commissioned officer at the drawbar tries to bob again, slurring a few sounds.

Reisiger feels that he can no longer control himself. He shouts, "If you make another sound, I'll bang you over the head."

Someone in the background says hazily, "Well, well, well!" But it remains quiet.

Reisiger looks at the sergeant. He sees that he is struggling to keep on his feet. "Sergeant, I hereby give you the official order to ensure that the battery is ready to march in one minute."

Someone laughs in the background.

Berger looks to the side. "Line up in two links, march, march!"

There is louder laughter.

The sergeant stretches out his arm. "With you, Herr Leutnant ... so ... orders from you ... if you say anything, the battery won't listen at all."

Reisiger, quivering with rage, says, "Sergeant, that's the end of it! Will you please tell me what this mess here means? You're an old corpsman, a soldier for fifteen years."

The red-haired man steps up next to the sergeant. "If Herr Leutnant wants to know ... we're not taking part anymore. We want to go home." And the sergeant continued in the same tone, "Besides, our first lieutenant is dead, and we can't get along with Lieutenant Berger."

From the crews squatting on the ground: "Bravo."

"Yes, and what are these new methods? Do you realize that you'll all be put up against the wall?"

Nobody answers. It is completely silent.

The circle of infantrymen closes in. Reisiger sees that they are already standing in three rows. Curious faces.

He turns back to the master sergeant. "Sergeant, do you realize what's going to happen now?"

The sergeant suddenly makes a helpless face. Runs his hand over his hair. Knocks his hand to the ground. "Then we'll just get ourselves shot to death .... It's all the same."

He turns away, sits down on the baseboard of the Protze, puts his hands in his trouser pockets.

The group of non-commissioned officers around him in silence.

Reisiger turns to Berger: "Herr Leutnant Berger, please have the battery's weapons collected."

Berger: "Listen up! All rifles and pistols will be collected here at the First Gun immediately!"

The redhead laughs. One of the squatters says loudly, "You're welcome to the weapons, but you have to get them."

An infantry officer has heard this grumpy sentence. He pushes his way through the men and goes to Reisiger: "Comrade, forgive me for interfering. I think it's necessary to set an example here. If I can be of any help to you . . . ." And before he gets an answer, he gives a command: "Second Platoon Regiment 58, fall in, in two links." His command shatters the circle of spectators.

The infantrymen jump backward into the forest. Seconds later, about thirty men line up in two links, field march style, rifles at the ready.

The infantry officer says proudly, "Please, comrade, dispose of us."

Dispose of us? Yes, how should Reisiger dispose of them? If only Captain Brett was here! "No one from the Eighth Battery is leaving this place . . . . You should be ashamed before your comrades . . . . But you'll have to pay for it."

Reisiger turns to Berger and the infantry officer: "I'll inform the captain. If the gentlemen will be patient until then."

8

And it may happen that you have to shoot or stab your own relatives and brothers. Then seal your loyalty with the sacrifice of your heart's blood . . . (or) . . . with the present socialist machinations it may happen that I order you to shoot down your own relatives, brothers, even parents—which God forbid—but even then you must obey My orders without grumbling . . . (or) *. . . and should I perhaps one day—God forbid—call you to shoot at your own relatives, brothers and sisters and parents, then remember your oath!*

(Wilhelm II, Nov. 23, 1891)

9

One hour after the incident in the Eighth Battery, several trucks appear on the orders of the regiment, on which the men are transported away. Nobody put up any resistance.

The next morning, the regiment decrees that the entire III Battalion is disbanded with immediate effect. The deputy squad leader and Lieutenant Berger are transferred home as fit for garrison duty. Officers and men are assigned to the batteries of the I and II Battalion. Lieutenant Schmidt 9/253 and Lieutenant Reisiger III/253 each take over one of the guns 9/253 for an anti-tank detachment, which is detached to the infantry by special order.

10

Reisiger reports to the regimental commander.

"Herr Leutnant, you have behaved scandalously. The gang should have been forced to obey at gunpoint. Do you know that you should be court-martialed?

But I'll give you one more chance. If I put you in the tank defense now, then you'd better prove that you're an officer...."

11

August 9   *Western Theater of War*

Penetrated our infantry lines with his armored cars . . .

August 11

Mass deployment of armored cars . . . more than forty destroyed armored cars in front of one division section alone . . .

August 12

Artillery closely following the armored cars . . .

Chapter VII

1

August 21　　　　　　　　　　　　　　　　*Western Theater of War*

Between the Oise and the Aisne, the enemy's renewed attempt to break through, which had been expected for several days and was initiated by strong attacks on August 18 and 19, began yesterday. After a heavy increase in fire, white and black French soldiers attacked in the early morning in deep formation, supported by numerous armored cars on a twenty-five-kilometer-wide front. They penetrated our front lines in places.

2

Fighting the enemy's armored cars, a new and hitherto little-known means of combat, was of particular importance in the assault defense. The destruction and barrage fire on hollows, paths, and enemy positions will probably bring the armored cars to a halt in many cases due to the mass of fire, so that only a few will reach and enter our lines.

These will be countered by guns (infantry guns or field guns) firing direct fire at close range. They are equipped with a special projectile for this purpose. It is important that these guns do not fire prematurely so that they are not detected and are still available if necessary.

Only such a tightly organized shelling of armored vehicles will be successful.

A general rule that all batteries that detect armored vehicles must fire against them will instead lead to confusion and failure.

(Combat regulations for artillery. Berlin 1917,
paragraph 299.
For official use only)

3

The Terrible Weapon

Before her Aisne maps displayed,
Waiting for the battle to be played,
Sat Queen France.
And around her the greats of the crown,
England's lords, with faces drawn.
"What would we be without,
Our tanks, so strong and stout!"

And as she waved her finger there,
To summon cute things in the air,
Steel giants full of storm and clang,
Indeed, the tanks.
"They'll," she laughed with a snap and shout,
"Crush Germany, there's no doubt,
For everywhere there's such a tank
In the ranks.
Hindenburg, I bet, would give a lot
To have such tanks, it's what he's not!
He envies us the thought, the tank
We've brought."

And to Sir Hindenburg, with sneer and jest,
Miss Marianne said with zest:
"Sir Knight, do you really believe,
That spirit can machines relieve?
Oh, so lift these tanks, if you please!"
And boom! boom! rang his cannons loud,
Crushing without pause or crowd
Twenty-six tanks in the field, so proud.

"Now he's got the tanks, our tanks indeed,
Now with them he'll surely succeed!"
But Hindenburg, with a smile, replied:

> "*Give me the sword that for Germany fights,*
> *The tank, madam, I don't need or seek!*"
>
> *And left them in the graveyard's muck and streak.*
>
> (Caliban, *Der Tag*, April 4, 1918)

4

August 22

The day passed quietly between the Somme and the Oise. Southwest of Noyon, we broke away from the enemy without a fight in the night from the 20th to the 21st. . . . Between Blerancourt and the Aisne, the enemy continued its attacks during the day. The enemy was only able to gain ground near Blerancourt.

5

The anti-tank guns of Reisiger and Schmidt were close to Blerancourt.

In addition to the two officers, the command included: Field Surgeon Winkel, two non-commissioned officers, ten gunners, and three telephone operators.

The two guns had taken up position during a quiet night without any casualties. They were positioned in an open field. A roof made of wire netting with piled-up leaves protected them from the view of airplanes.

Firing the guns was expressly forbidden. The only purpose was to achieve a surprise effect against armored cars in direct fire.

Officers and men lay in readiness in a spacious limestone cave. They shared the quarters with the reserve of an infantry battalion. The leader of this reserve, Major Sänger, also took command of the guns as section commander.

The cave was the kind of quarters they had not had for many weeks. It consisted of a huge room the size of an unusually large station hall (it was occupied by the infantry) and a smaller vault, no larger than five rooms of a private apartment. This was where the tank defense command was located.

Both rooms were connected by an underground corridor. There was no need to go outside for days on end. You were guaranteed to be out of sight of the planes. You could even cook in the corridor.

The cave for the artillerymen must have been inhabited before they moved in. There was a table, a few boxes, and, on one wall, stacked racks of straw: beds unlike any seen for a long time.

The single room provided a new community. In the artillery, officers and enlisted men had rarely lived together like this. They cooked together, ate together. Toward evening, when a second candle was lavishly added to the one that was constantly burning during the day, everyone sat at the same table.

There was no mail, no newspapers; the less one was connected to the outside world, the sooner one's tongue loosened.

What was said?

Never much. But always, after just a few sentences, the topic was the same: When will the war end?

Reisiger often found himself dying to hear more from his fellow diners; the answer to the question: How will the war end?

Sometimes the answer came. Nobody talked about victory. There was more talk of negotiations with the enemy, of liquidation. And otherwise, not much more thought than the primitive: we've all had enough. We want to go home . . . That had already taken root. That was already the law.

But how do we get home? Who decides that? Who makes the end? No, no, they didn't ask that.

And by the way, we lived quite well. We were fed by the infantry. Plenty of dried vegetables, some bread, sparse jam, and very little cheese. A "cheesecake" was invented; slices of bread as thick as a finger were spread with jam and cubes of cheese were placed on top. The whole thing was held on the side of the fire until the topping melted.

6

French airmen dropped small white notes. It was forbidden to read them, and orders were given to hand them over to the nearest officer as quickly as possible. They were read anyway; they were not given to the next officer, but to the next comrade. And they read them day after day:

German comrades! Tell me, when are we going to put an end to this eternal killing in the interests of the Hohenzollern[22] megalomania and a few big moneybags? Because every one of you knows as well as I do why this war is happening, independent from which German state you are. Do not be so afraid, comrades, and remember that we have the power in our hands, that the power is ours. The Prussian, trusting in your servant spirit and cowardice, has even given you a shotgun and 150 bullets to fight under the motto: With God for King and Fatherland, to ensure that the chains of slavery are forged even tighter on you and your descendants than they were before. So, wake up, German Michel, and break the chains, become a man and set yourself free, so that your descendants will not despise and mock you. I too am married and have children, and yet I have made myself free, for what woman can still respect a man who after four years is still fighting for his slave masters, who is not afraid of death during an attack and yet does not have the courage to suffer death for his rights, his freedom and the liberation of his loved ones? If you do not have the courage, comrades, come over to beautiful France; here you will not only be fed and have good food, but you will also be treated like a human being and not like a human-like animal.

A comrade who feels like a newborn human being here!
    Pass it on!

Back cover:

It generally seems as if the world war should be seen as a family affair for the Hohenzollerns. "Wilhelm has attacked," read a recent army report. Wilhelm did not attack, but thousands of soldiers had to attack while Wilhelm was forty or sixty kilometers behind the front. Better peace without a monarchy than war with a monarchy.

(Dr. Cohn, MoP, U.S., Reichstag session of June 14, 1918)

---

[22] The Hohenzollern dynasty. Founded before 1061, ruled over the kingdom of Prussia since 1701, and the German Empire from 1871 to 1918. Kaiser Wilhelm II was its last ruler.

7

Reisiger and Schmidt reported to the infantry major every evening after dark. There, they read the army report. It was incomprehensible how bad it looked everywhere, how the machines of the front lines were crashing into each other. Incomprehensible: Because there was absolute peace here.

"We'll end up sleeping through the whole war here." The major growled. "My people are losing all combat effectiveness here."

"But, Herr Major, what is not yet can still be."

"If we're lucky . . . that our terrain doesn't encourage the enemy to attack . . . You see . . ." War was played out on the map. Again, and again, all the chances were thought through. "Yes . . . and tanks . . . ?"

Nobody knew tanks. Nothing like that had ever appeared in this section before. Looking at the map, it was also difficult to work out where the beasts were supposed to have their stables. The terrain is flat, rising gently. And on this side of the height, six hundred meters in front of it, our infantry is lying in the craters. What do they mean with tanks?

"Presumably, the enemy has massed artillery and armored cars south of us. And we're looking at the moon. A shitty war. What do I have my battalion for?"

So—what else can you do—we play skat.

Every evening, until around midnight.

8

Reisiger comes with Schmidt from the major.

Everything is asleep in the cave. Gorgas, the telephonist, is on the phone. The line to the regiment is in order; there is nothing new. Reisiger is still hungry. He digs some bread out of the sandbag next to the bedstead and sits down at the table. There is a dried-up piece of sausage wrapped in a newspaper. His eyes read the newspaper.

He is too lazy to speak. He can't take his eyes off the newspaper either.

Schmidt is interested: "What are you reading that's so important?"

Reisiger grumbles distraught: "Oh, nothing special."

Schmidt realizes that there's no point in encouraging Reisiger to have a conversation. "Well then, good night." Skirt and boots off, off to bed.

Reisiger is still reading. He doesn't know what he's reading. It doesn't enter his consciousness. It is only spelled out. What does that mean? "Tell me, Gorgas, is this your sausage?"

"Yes, if Herr Leutnant wants to help himself."

"No, thank you, I'm only interested in the paper." And he unwraps the sausage, puts it aside, and flattens the paper. So, news from home. That's still quite nice. They really don't feel anything about war. Fat life.

It's almost as if you've forgotten how to read during the long postal embargo. You have to read each letter individually, slowly putting each word together. Half aloud:

When the war broke out, we thought it would be a short war, but things turned out differently. Russia's declaration of war was followed by France's, and when the British also attacked us, I said I was happy about it, and I was happy because we could now settle accounts with our enemies and because we finally had direct access from the Rhein to the sea. Ten months have passed since then . . .

And I am happy about that, and I am happy about that, and I am happy—ten months—ten . . .

Reisiger wants to turn to Gorgas: Have you heard that King Ludwig has made a speech? But . . . "ten months" . . . and he looks at the report again: June 6, 1915.

Is that so? On June 6, 1915, King Ludwig rejoiced over the enemy. Is that so? "Much enemy, much honor." Is that so?

He wraps the sausage up again, tightly in the crumpled sheet. He pushes it across the table, far away from him.

Oh well, times have changed. And where was the joy?

June 6, 1915, can you still think that today?

"Herr Leutnant, I've just overheard a conversation from the infantry that the enemy attacked south of us in the evening. He has been beaten off. The telephonist said that there were mountains of corpses lying in front of the trenches . . . . This is almost no longer a war . . ."

It's almost not war anymore? Yes, it is war! "Gorgas, when will you be relieved?"

"At two o'clock, Herr Leutnant."

"I'm going to bed now too. If anything happens, wake me up."

Reisiger can't find his canvas. He has the telephone operator give him a candle and lights up the individual beds. Maybe someone has stolen it.

As he approaches the sleeping people, he forgets all about searching.

He remembers the night before Arras when he saw the first dead. Those obliterated white faces.

Now it's here again. Now it's so strong that you can't decide: Are the people lying there stretched out between the filthy canvas alive or not?

These faces. Like burnt-out craters. With very sharp walls. The nose is steep and narrow, the sharp cheekbones, the eye sockets deeply depressed, the mouth pinched, or slightly open and slack. And all white, much whiter than the lime.

That is incomprehensible.

Why does it have to be like this?

The old Gunner Dietrich lets one hand hang to the ground. A thick, wrinkled hand with short fingers. On one of them, a particularly broad, reddish wedding ring.

Why is Gunner Dietrich lying here?

Next to him is the doctor, Dr. Winkel. He's even more frightening to look at. He's completely decayed. Probably already dead. Why does he have to be dead?

Why?

The two sergeants are lying next to each other. One of them has put his hand under the shoulder of the man next to him. Now this hand comes out from behind his neck. Maybe it's about to wrap itself around his throat.

But no, it's also white and stiff.

If only someone could understand why?

Reisiger slowly walks back to the table. He knows that back there, even further into the black of the cave, there are more! There lies Schmidt, and there lie the two young gunners, the nineteen-year-olds who only recently arrived. He doesn't want to see that anymore. It all just screams Why? Why, why, why!

The telephone buzzes. The telephone operator picks up the phone, listens for a moment, then says loudly, "Why?"

What's going on now?

Something is tugging at Reisiger's nerves.

He drops the candle, quickly steps next to Gorgas, and shouts at him, "Why are you saying 'why'?"

Gorgas is startled. He stands up, makes an astonished face, stammers, "Why do I say 'why?' The regiment called to see if we needed any more ammunition . . . And then I asked, Herr Leutnant, will forgive me, why?"

He looks at Reisiger with helpless eyes.

Can't you ask why?

Reisiger pulls a crushed cigarette out of his coat pocket. He breaks it in two and gives one half to Gorgas. He put a hand on his shoulder: "Sit down again. I just wanted to know what's going on."

They smoke. It takes four puffs, then the cigarette is finished.

Reisiger turns away again, back to his bed.

But then he goes to the exit of the cave and steps outside.

A starry sky. The two guards are crouching above the entrance, whistling. There are occasional flashes of light at the front. But there is no gunfire in the area.

There is the Great Bear. Cassiopeia. The big W.

The big W.

Reisiger plays with the sound on his lips. The big . . . W . . . the big W . . . W . . . . Breaks off: Reisiger, you're crazy.

He is freezing. He wants to go back into the cave. The big . . . W, big . . . W . . . ; he laughs: W means? The big W, . . . big W, . . . big . . . that is?

"Gorgas, do you have another cigarette?"

Gorgas rummages in all his pockets and searches the lining of his cap. "No, Herr Leutnant, nothing left."

The big one—the big one.

Of course. There's only one capital W. This nonsensical, this nonsensical: Why. W . . . why?

And it stands here above us and above all the trenches and above all the battery positions and above all the staff quarters and above all the high commands; there stands the W, stands the W . . . why?

This is a discovery that makes Reisiger happy. He would like to wake the doctor and tell him. Schmidt wouldn't understand. He's still too young. Why would he ask why?

But it's an old rule that you shouldn't wake anyone out here if it's not necessary. Let him sleep.

Where the hell is the canvas? All this thinking wouldn't have happened if the canvas was in the bed.

But that doesn't matter. He puts his skirt over his knees and keeps his boots on. Reisiger falls asleep.

He can still hear Gorgas being relieved by Trebbin. So, it's two o'clock.

9

Three hours later. Five o'clock in the morning. The two guards at the entrance to the cave are lying on their stomachs. They are dozing. In an hour, they will be called in.

Then, as they both look at the edge of the horizon toward the enemy, the earth moves. Blackish blocks rise up against the pale gray of the sky in six or eight places along the front. Or blackish mountains. Or houses with gables taller than a man, slowly rising.

The two guards have never seen a tank in their lives.

They know: These are tanks.

Where the black blocks are crawling now, a few seconds after they appear, that's our infantry front line.

Why isn't the infantry firing?

There is actually only one order for the guards of the tank defense platoon as soon as a tank becomes visible: "To the guns."

The two soldiers reach into the soil and stare ahead.

One there, a tank, another one there, five, eight, eleven, two more there.

It's incomprehensible.

No, it's comprehensible: The enemy begins a fire attack. It must have been opened simultaneously to the second by several dozen batteries. It whistles, roars, claps, and hails.

One of the first heavy shots hits the entrance to the cave.

The two guards fly up in an arc, lie flat, and jump into the entrance. Shouting, "To the guns!"

Everything in the cave is confused. The two guards stand in front of Reisiger: "Herr Leutnant, they're here."

A stream of fire hits the entrance. The telephone operator almost flies into the back wall with his device. The table rears up.

Everything is awake. But thinking stops for a moment. "Who's there?" Reisiger looks at the guard and turns to Lieutenant Schmidt. "Aha ... quite a decent barrage."

"Herr Leutnant ... tanks!"

Tanks? Yes, is that word alone paralyzing? Reisiger has to give himself a jolt. But he sees that everyone else is still standing there frozen.

He shouts, "What are you standing there ... you hear me ... tanks ... Herr Leutnant Schmidt ... quickly, to the guns!"

He takes his steel helmet and gas mask and goes to the entrance.

The view is blacker than the wall of the cave. Drumfire. You can no longer make out individual shots. Just a black curtain, twitching, higher, lower for a moment, then higher again. Flames in between.

Everyone has now stepped behind Reisiger.

Schmidt: "Damned coffeehouse—how are we supposed to get to our guns?"

Reisiger: "One by one ... jump, march, march ... whoever is lying down stays lying down ... do we have contact with the regiment?"

The buzzer sound is barely audible. "No, Herr Leutnant, everything is broken."

"And the infantry? Schmidt, why don't you walk through the corridor ... what's the major doing?"

Schmidt off.

"Sergeant Gross, you go first."

"By your command, Herr Leutnant." And smiling a little. "If that goes well ... Lehmann, you come next ... about ten paces away ... and then the others."

Another shot hits close to the cave. It was right on the roof in front of the entrance. A wall slams inwards.

Reisiger pushes the dirt aside a little with his foot: "The better the cover will be ... so let's go Gross."

Shot, the fire splashes hot against the walls. Gross jumps off. He has disappeared in the black smoke.

"Two ... three ... four ..." Reisiger counts out loud to ten. "The next one!"

And the next one is swallowed up by the black wall.

Where is Schmidt? What is the infantry doing?

You can hear footsteps through the connecting corridor: "Reisiger, the cave is empty. The infantry is probably in the rear position . . . ."

At least according to the footprints. They've only left a few wounded behind."

Nice prospects. If the first line is overrun now, we'll be without cover.

Suddenly, a roar in front. Lehmann, the gunner, bursts out of a column of fire. He is on his knees, without a steel helmet, dragging something behind him. He straightens up, reaches backward with his other hand. He lays the sergeant Gross at Reisiger's feet. One arm is off. The uniform coat is torn open from the neck to the hips. Gasps: "Herr Leutnant, there's no chance of getting through."

Reisiger looks at the sergeant. How he bleeds. Two gunners drag him off. The doctor arrives. "There is nothing more to do. Thank God he was killed instantly."

Another shot at the cover. The entrance narrows. Everything steps back a little more.

"Herr Reisiger... then I'll try it first." Lieutenant Schmidt tightens the storm strap on his helmet.

"I'm afraid, dear Schmidt, we all have to, whether we like it or not. All we need is for the enemy to shoot up the entrance. Then we'll suffocate here. Because the infantry exit is right in the direction of fire .... So, I go first. We'll all follow at a distance of ten paces; boys, as soon as we get to the guns, nothing can happen to us."

Reisiger jumps off without another word.

Now Schmidt is counting.

The first gunner is gone.

The second gunner is gone.

Now comes Sergeant Dircksen. Gunner of Reisiger's gun. "Nine, ten .... off!"

Just as the sergeant is about to disappear, he jumps up and strikes backward. Dead.

Now the next one. Schmidt is unsure: "Well, who's next?" No answer.

So, first the two boys. "Ziese... out with you."

Gunner Ziese clicks his heels together, as he learned on the parade ground.

Then he kneels down, a runner before the start. Off! Jumps over the dead sergeant, just out of sight.

The second of the boys cries out, "Yes!"

Then Ziese also falls backward.

Schmidt is at a loss.

The black wall dances up and down. Now, for once, the whole cave rises. You can clearly feel that the floor is pushing upward.

Jesus Christ Sacrament—we have to get to the guns!

New shot! Reisiger stands there as the smoke drifts to the side. He stumbles back into the entrance. "It's no use, both guns are in pieces. Load the carbines, the enemy is right here."

Load carbines? Artillerymen know as much about carbines as infantrymen know about cannons.

Everything is confused.

Reisiger, senseless: "The enemy is right here, don't you understand me?"

"Herr Reisiger, what do we few men actually want with the enemy?" Schmidt asks somewhat mockingly.

Reisiger looks at him angrily.

Schmidt corrects himself. "Of course we're going to defend ourselves until the enemy stops shooting."

They all push toward the entrance again. No, the enemy artillery is no longer firing.

Instead, there is a barrage of gunfire.

Gunfire makes no impression. Just gunfire. Everyone gains new courage.

"Above all, we'll find out what's going on," says Reisiger. "Come on, Schmidt."

Both take carbines and walk out of the cave. Upright. But then they quickly drop to their knees and lie down.

"Schmidt ... there they ... are!"

The noise of the infantry fire is drowned out by a roar never heard before.

A hundred meters in front of Reisiger, a tank crawls up.

A giant from the primeval world.

A monstrous, blackish, naked back, carried ponderously by two enormous thousand feet that shuffle and grind over the earth.

Two nostrils on its head bark incessantly and hiss incessantly yellowish, stinging flames.

A tank! And not far away, a second beast. And already a third, a fourth.

And barking and hissing and shuffling closer.

There stands a tree, tattered, pathetic, but at least its trunk is still stretched toward the sky. Stop—and the tank goes for it. The trunk collapses to its knees. It is crushed.

There are bushes, wire entanglements, concreted deep into the earth. Stop! The beast pushes through without pause.

Where is our infantry? Where is our first line? "Schmidt, where is our first line? There must be infantrymen somewhere!"

Schmidt straightens up a little, supports himself with his arm, points. "There are . . ."

Reisiger hears something hit hard, like stone on stone. The shot is right between Schmidt's eyes. He falls over. "Schmidt!" No time. "Schmidt . . . there are?"

Reisiger straightens up. *What did Schmidt see? There are . . . there are no German troops!* He crouches lower. *There are Americans.*

Machine gun fire. Everywhere. Even from behind. Thank God, it's our infantry in the second line.

Reisiger looks around. Maybe it's the major with the reserves.

But in front of him, and this is more important, and this is threatening: the tanks, closer and closer, closer and closer. And half right, yes, you can already see it clearly: American infantry.

Rifles!

Schmidt is dead. *What am I supposed to do as a single person with a rifle?*

Hopefully, our men in the second line will take care of that on their own.

Back to the cave: We'll defend ourselves there.

Back to the cave.

"Winkel, Lieutenant Schmidt is dead . . . Boys, now it's about the last bit. A tank will be here soon. The enemy has American infantry . . . We have to fight our way backwards."

Reisiger looks at the men. They're standing there, hunched over. Sure, the two sergeants are dead, Lieutenant Schmidt is dead, and the two war volunteers are dead. But you're used to that sort of thing. "Or what were you thinking?"

They are thinking something. You can see from their faces that they're thinking something. "So speak up . . . every minute is important . . . Well, Gorgas, what do you think?"

Gorgas steps from one leg to the other, looks at Reisiger and Dr. Winkel and then at his comrades. "Herr Leutnant, I mean ... we were just thinking ... earlier, while Herr Leutnant was outside, we brought the wounded to us from the infantry in the big cave at the back. And then Herr Doktor said ... and I mean ... we put them here so they wouldn't be alone. And are we not allowed to look after them?"

Reisiger doesn't quite understand. He looks at Winkel. He shrugs his shoulders. Is that approval or disapproval? "So, what's going on now?"

Gorgas continues, "And that's what I meant ... if the enemy comes and we say that there are wounded here and that we're something like medics ..." He says no more.

"So you want to be taken prisoner?"

Silence.

"Guys, you must have been abandoned by the good Lord. I can't just give you seven men here to the Americans. Besides, do you realize that they won't wait for you to be good enough to come out of the cave to meet them? You know they'll throw a hand grenade into every hole. Would you rather have that in your face?"

Then the infantryman Dietrich steps forward. "Herr Leutnant, if you don't mind, I'd rather have it this way ... and then it's really because of the wounded."

What to do?

Reisiger goes back to the exit. Yes, that strange sound of the tanks rolling in has become stronger. Perhaps it really is only a matter of minutes. Anyone who is not out of the cave by then will have to stay here.

He places his steel helmet firmly on his forehead: "Doctor, are you coming with me?"

He wanted to say it as a question. And now it has become like an order. Winkel swallows. "Of course."

"So, you want to leave me alone?"

Gorgas: "Herr Leutnant, it's because of the wounded."

"Then please put a white handkerchief at the entrance. And whoever is caught, write me a postcard. Come on, Doctor."

10

The first fifty meters in a crouched position. Winkel behind Reisiger. Damn it, the enemy is still firing artillery. The shot was right in front of my nose. So, lie down. The splinters whiz up singing. They don't hurt you anymore. Look around: the poor guys in the cave! The one tank must be with them in a few minutes.

Go on. Crawl on your stomach. It's difficult because the path is trampled. It's always up and down, head up, then back on your feet. Now another twenty meters, then there is a little cover, a miserable hollow path. With a wall, shoulder-high, to the rear. With cover, little more than twenty centimeters to the front. Better than nothing. If only the enemy doesn't spot you.

That's how they crawl, crawl. Artillery shots. The machine guns bark. Since they drag their noses in the dirt as much as possible, they can't see much. Reisiger becomes reckless. Oh no, three steps, then the hollow path is there. One jump, maybe two meters. Cover! Winkel behind.

Winkel pushes forward; they are head-to-head.

"A funny thing. I feel like we're the only two people on the whole of French soil. Except for the tanks, of course. But they won't hurt us. I must have seen some Americans earlier."

Both raise their heads, cautiously. Now they are out of cover. But there is only a narrow gap between the steel helmet and the ground.

"Do you see anything, Winkel?"

And they both pull their heads back in at the same time. Good heavens, there aren't a few Americans. They are many, swarming around. Maybe two hundred meters away. "Winkel, they're walking upright as if they were taking a stroll!"

Winkel doesn't answer.

Behind them, furious machine gun fire. Reisiger slowly rolls onto his back, maybe you can see who's actually shooting. Oh yes, you can see it.

You can see very clearly that the shoulder-high cover goes up centimeter by centimeter in small, spraying fountains of sand: The German M.G.s of the rear position are combing this very slope.

"We have to crawl on our bellies."

"But Herr Reisiger, it could go on like this for hours."

"Yes, should we jump?"

That might work. Reisiger crouches down. At the same moment, there is fire from the front. As he slides back onto his stomach and looks up, the same small fountains that are bouncing back and forth on the rear cover are now also bouncing in front of them.

That means: They are right in the line of fire of the machine guns of friend and foe.

And that means being pinned down here for no purpose.

The noise of the tanks gets louder; the small fountains bounce more vividly. The Americans are probably getting closer step by step. So, the end will be: getting hit on the head with a piston. "Yes, dear doctor, there's no other solution; we have to get out of here."

Winkel, grimly, laughing: "That seems to me to be the case."

How well you have your thoughts together. Reisiger lectures in a loud voice: "It's a shot to the head or a shot to the stomach. You, dear doctor, will agree with me that the head shot is the more pleasing one. So, if we stand upright, our own infantry will shoot us in the head. But if we bend down and run with our knees bent, the Tommy will hit us in the stomach. Ergo: we walk upright."

One, two, three, they stand. They start to move, upright, or as good as upright, with their heads carefully tucked on their chests. Look to the right; there is the high embankment, yes, and there, always at shoulder height, the German bullets hit.

Look to the left, look to the left, and then they forget all premeditation and deliberation. There come three or four rows, more upright than them, hooked underneath, their rifles casually tucked under their arms as if on a chase, Americans. So close that you can recognize their faces. So close that you can see individual groups laughing, and swaying back and forth as if they were walking in a dance step. There on the left, a separate group is kicking a football in front of them.

Again and again, one and then another suddenly throw up their arms and slump down. The ranks close. The living walk on.

Covered by nothing but the tanks.

And the tanks, they shuffle closer. Their roar gets louder and louder, duller and duller, more and more surreal.

A few seconds of these images. Then they shuffle off again!

Reisiger and Winkel see that the chain of small fountains follows them at their feet and always follows where they run, ten centimeters beside them, twenty, closer, further, and always prancing.

They plunge into a gallop. Further along the hollow path. Through shell craters. They sink up to their shoulders, climb out again, and plunge headlong into new craters.

The enemy is getting closer.

And the tanks are firing. And the Americans shoot in front of them, and the Germans shoot behind them. And the barrage becomes a fortissimo.

Run, run! It's an insane heat; it must be midday. The sun is burning.

Running. As Reisiger takes a look around, Winkel throws off his skirt. Continues in his shirtsleeves.

The embankment ends. It doesn't help. Up into the open field. Just get away from the enemy!

Reisiger jumps onto the plateau first.

He sees that the German infantry is just climbing out of the reserve position. "Winkel, rescue is coming!"

But no. But madness! There is no assault and no defense. A tank has already pushed through to the second position, oh, far through. It is now driving directly along the line of the trench and firing two M.G.s directly into the cover. And the infantry jumps out, runs away, run, run, save yourself if you can.

So after them!

One look around ... tanks everywhere. Everywhere is the quadruple wall of the Americans, step by step.

The terrain rises. The higher Reisiger and Winkel get, the better the view, the more clearly they see the cauldron closing in.

Columns upright, Americans, English, and French. And tanks in front of them.

And in the cauldron, thousands of people, the Germans.

Everything is running, running. There is still one open spot in the jam: "Towards Germany! Back!"

And again and again, wherever you look: this strange movement, the slight jump, almost straight up in the air, and arms screaming for help, and sinking together.

Running, running.

Battery position. Everything is dead. The guns are scattered, forming inextricable piles of rubble.

Machine gun emplacements. But the crews are gone.

And dead, dead, and wounded everywhere.

Run, keep running!

In the meantime, the enemy has brought its artillery forward. The barrage begins in the rear area. After ten minutes, the whole area is covered in columns of fire as thick as a man and as high as a house.

The Germans have to get in there!

And the tanks are coming. And behind them, Americans, English, and French.

There's a forest there. It is burning. But it is alive. Through the lanes of its trunks, the infantry is rolling backward in thick columns.

There is a highway; columns of wagons and horses and guns and cars are heading backward along it.

Above them, the sky darkens. From a height of a few meters, flocks of planes swoop down on them. Machine guns mow life to the ground.

The columns come to a standstill, overrun each other.

New flocks of planes drop chain bombs. Bombs that produce a hundredfold explosion, kill and tear and kill a thousand times over.

Run, run, run.

11

7,057,000 men against 2,500,000 men.

12

The national defense, the uprising of the people must be initiated, a defense office must be established. Both will only come into force when necessity demands it, when we are pushed back; but not a day must be lost.

The office is not to be attached to any existing authority; it consists of citizens and soldiers and has broad powers.

Its task is threefold.

Firstly, it appeals to the people in a language of unreservedness and truth. Anyone who feels called upon may come forward; there are plenty of older men who are healthy and willing to help tired brothers at the front with body and soul.

Secondly, all the men in field gray who can be seen today in towns, at stations, and on railroads must return to the front, even if it may be hard for some to interrupt their hard-earned leave.

Thirdly, in East and West, in stages and in the hinterland, those carrying weapons must be weeded out of chanceries, guardrooms, and troop posts. What use to us today are occupation and expedition forces in Russia? At the moment, just more than half of our troops are on the Western Front.

A renewed front will be offered under different conditions than a tired one.

Walter Rathenau,[23] *Vossische Zeitung*, Berlin, October 7, 1918

13

Under all circumstances, the impression must be avoided that our move towards peace comes from the military. The Reich Chancellor and the government have taken it upon themselves to take the step on their own initiative. The press must not destroy this impression. It must emphasize again and again that it is the government, true to its repeatedly expressed principles, that decided to take the peace step.

Press conference, October 16, 1918

14

General Ludendorff has just asked Baron von Grünau and me, in the presence of Colonel Heye, to convey to Your Excellency his urgent request that our peace offer be sent out immediately. Today, the troops are holding up; what will happen tomorrow cannot be foreseen.

(von Lersner, representative of the Foreign Office, January 10, 1918, one o'clock afternoon)

---

[23] Walter Rathenau. 1867–1922. Leading Jewish industrialist in the late German Empire. Played a key role in the organization of the German war economy and headed the War Raw Materials Department from August 1914 to March 1915. Assassinated by nationalists in 1922.

15

Since Reisiger, as he is found and taken to the General Command, declares that he considers the war to be the greatest of all crimes; he is arrested and locked up in a madhouse.

16

Reisiger lies in an isolation cell. It is a tomb, dark, cold, lit by a bluish lamp. The door is locked; the window barred with centimeter-thick glass.

So, now I'm buried. Now it's over. Now it would be necessary to write to my mother that I am lying here. But no one would allow that. I am crazy. I am mad on the highest orders of the highest commanding general. That's the way it has to be. An officer who runs away, who no longer plays along, is crazy. Running away is crazy, running crazy, running ... It's hilarious how I'm lying here. And I didn't even tell him that I wasn't playing anymore. Herr General, I only said, please shoot me, here, help yourself, but I won't take one more step forward. I will no longer commit the greatest crime of all ... Where have you been for so long? And why don't you stop the tanks, eh?

And, Herr, you should tone it down, he said. And I shouted that the hussar with the patent-leather boots had turned pale; I don't think of moderating myself, I said. I've been moderating myself for far too long, and if I hadn't moderated myself earlier, all those who fell would still be alive. I say as loud as you want to hear that we are all complicit in this senseless crime and I won't tolerate anyone laughing here now, and besides, count on it, the tanks are coming right here to the village—and grabbed me—why didn't I fight back ... and put me in the car, strapped to the stretcher, and pushed me under the bench on which a man was bleeding to death without legs, so that my face was wet. And yesterday, driving through the city, in the barred car, laughing and singing and declaring, with all my fervor, to all the doctors: Gentlemen, I swear to you, I'm not crazy. Nor am I acting crazy.

I declare to you with my life: I know what I'm doing, and I say, it's about nothing other than saying, I, I, I'm not taking part in the war anymore. I'm not taking part in the war anymore. I know I'm letting my comrades down, and maybe that's cowardly. But yes, I am a coward. I want to be a coward. I keep suggesting it to you: Why don't you shoot me? Why don't you impose your ridiculous laws of war on me and shoot me? But I won't take part anymore. I no longer want to be complicit. It's about more than victory, which you believe in just as little as I do. It's about the fact that people are still being shot, beaten, and mutilated every second—and for what? For the sake of futility, because we can no longer win. We have fought out there for years like no army in the world; we have had all faith, even when we said no. Now it's enough. And I'm not in it anymore. And I'm not taking part anymore.

But then you laugh and pity me. "Take your hand off my forehead," I shout at the doctor. I don't want to be comforted. I'm not to be pitied, I'm not ill, I'm not mad, I don't want to be excused, I'm telling you, I know what I'm doing. War is the greatest crime I know. I am to blame for it. I have been guilty of it for years. People have been killed by my command. Now it's over. Let me pay for it. Why don't you kill me, because I knowingly, knowingly let you down . . .

But as I cry, you laugh even more pitifully, saying, poor, crazy lieutenant. And I am clearer than ever before in my life; it is a crime to take part in the murder for even a second longer.

17

Mainz fortress hospital, mental ward. Weekly report September 6–13, 1918. orderly: Neuhagen.

Reisiger, Adolf, Ltn. d. R. F.A.R. 253. Findings, as in the previous week, the patient does not sleep, does not eat, looks stolidly in front of him. When you talk to him, he always has only one sentence to answer: "It's still war. Kiss my ass!"

Epilogue

There fell in the years fourteen to eighteen:

One million eight hundred and eight thousand five hundred and forty-five Germans, one million three hundred and fifty-four thousand Frenchmen, nine hundred and eight thousand three hundred and seventy-one Englishmen, six hundred thousand Italians, one hundred and fifteen thousand Belgians, one hundred and fifty-nine thousand Romanians, six hundred and ninety thousand Serbs, sixty-five thousand Bulgarians, two million five hundred thousand Russians and Poles, fifty-five thousand six hundred and eighteen Americans.

Together: Eight million two hundred and fifty-five thousand five hundred and thirty-four people.

Afterword

Jens Malte Fischer

Deutsche Verlags-Anstalt 2004

The new edition of Edlef Köppen's *Heeresbericht* (1930) makes available once again the major work of an author who the general public has largely and undeservedly forgotten. This is due, among other things, to the early death of Edlef Köppen, who died in 1939 at the age of just under forty-six, but also to the National Socialist cultural policy, which wanted to erase the work of this author from literary memory. His *Heeresbericht* was banned in 1933; in May 1935, the last copies were confiscated by the police president in Leipzig, and the book was finally added to the "List of Harmful and Unwanted Literature" in December 1938. The continuous reception of Köppen's works was prevented during the Nazi era, and after 1945 it did not initially resume. Interest was reawakened in the 1970s, particularly in the novel *Heeresbericht*; there were even several new editions. However, this revival also petered out in the early 1990s. Those who are interested in the literary treatment of the First World War and who may previously have only known Erich Maria Remarque's incomparably more successful novel *All Quiet on the Western Front* (as a newspaper preprint in 1928, as a book edition in 1929) will find here a comparable moral stance, but a more advanced literary form and a far more precise depiction of the horrors of war in terms of facts and atmosphere.

Edlef Köppen was born on March 1, 1893, in Genthin (Mark Brandenburg, now Saxony-Anhalt) on the Elbe-Havel Canal, the son of a general practitioner. Köppen's response to the slander of the *Völkischer Beobachter* (the newspaper of the Nazi Party (NSDAP)) of August 1932, that he was Jewish and had only recently acquired his naturalization, shows that on his father's side, he came from a farming family that had been resident in the Altmark for 300 years. His mother came from a Holstein merchant family. It was already the time when one was forced to point out one's 'Aryan' descent, even if one was as free of anti-Semitism as Köppen. At the age of fourteen, Köppen moved with his family to Potsdam. At the local humanistic Viktoria-Gymnasium, he met Hermann Kasack, who was three years younger. Together, they were active in the student association "Literarischer Abend" (Literary Evening), where they discussed contemporary literature. In 1913, the year he graduated from school, Köppen began his literary work, according to his own account. He began studying German in his mother's hometown of Kiel and continued this after a year at the University of Munich, while also attending courses in art history and philosophy. His most influential teachers were Fritz Strich, Heinrich Wölfflin, and Artur Kutscher, who later founded the Munich School of Theater Studies. Kutscher and his popular excursions were even mentioned in the *Heeresbericht* and attest to the novel's highly autobiographical nature: In one scene, the protagonist Adolf Reisiger and his superior, Lieutenant Stiller, recall a peaceful moment in peacetime, the lectures at Kutscher's, and a trip to Salzburg together. Köppen had access to the legendary author evenings that Artur Kutscher organized, where people not only discussed and drank, but also read from their own manuscripts. Max Halbe, Rudolf G. Binding, and Joachim Ringelnatz were among the earliest guests. In 1911, Frank Wedekind, Hanns Johst, and Alfred Henschke (Klabund) joined them, followed in 1913 by Josef Wenter and Josef Magnus Wehner, who later wrote the nationalist war novel "Sieben vor Verdun." And it was Kutscher who remained vividly in the memory of the young Köppen when he shook his hand one night in front of the Munich Kammerspiele in the summer of 1914 and said: "Now show that you have not only learned theories from me, and do not forget that all great poets were also decent guys." These words were addressed to the war volunteer Edlef Köppen, who reported to the 40th Field Artillery Regiment in Burg (near Magdeburg) in mid-August to complete his basic training before going to the front.

Köppen was one of the few writers who fought in the First World War as an active soldier from its beginning to its end—Remarque, who was five years younger, served "only" from 1916 to 1918. In October 1914, Köppen was transferred to the Western Front in France (probably near Arras) as a gunner in the first replacement division of the 40th Field Artillery Regiment. He took part in the Battle of Loretto and the fighting near Souchez and Loos, was twice treated in the military hospital in Douai and finally fought in the Battle of the Somme in 1916, during which he suffered a bruised lung and was admitted to the military hospital in Genrode (Harz Mountains in Saxony-Anhalt) for two months. According to family accounts, this injury was a major contributing factor to his relatively early death, a good twenty years later. At the end of 1916, Köppen's regiment was transferred to Russia, and in the spring of 1918, it was brought back to the Western Front. Köppen experienced the end of the war in a psychiatric hospital in Mainz (Rhineland-Palatinate), after he had resolved the conflict between fulfilling his duty and recognizing the immorality of the war by openly refusing to obey orders. In December 1918, he was discharged from military service as a lieutenant in the reserve and holder of the Iron Cross, First Class. The confusion at the end of the war prevented him from having to bear the consequences of his actions. After the war, the health problems resulting from his war wounds led to repeated biographical career breaks and to changing activities in the field of literature.

Edlef Köppen resumed his studies of German language and literature at the University of Munich in the summer semester of 1919, after he had completed only three semesters (1913 to 1914) before the outbreak of the war. He did not finish his studies, despite having almost completed his dissertation on magazines of the Romantic period. On December 1, 1920, Köppen began his career in publishing at Gustav Kiepenheuer Verlag in Potsdam. His school friend Hermann Kasack was an editor there, and Hedwig (Hete) Witt, whom he had met in 1917 during a home leave and was to marry in 1921, was also employed at the publishing house at the same time.

Founded in 1909 in Weimar by the bookseller Gustav Kiepenheuer, the publishing house developed into one of the most prominent in the field of contemporary literature thanks to the prudent owner and his outstanding editors Ludwig Rubiner, Hermann Kasack, and Hermann Kesten. Leonhard Frank, Upton Sinclair, Ernst Toller, Anna Seghers, Joseph Roth, Gottfried Benn, Heinrich Mann, Lion Feuchtwanger, Joachim Ringelnatz, Hans Henny Jahnn, Georg Kaiser, and Arnold Zweig were among the authors. Zweig's novel *The Dispute about Sergeant Grischa* (1927) became the most successful title in the history of Kiepenheuer Verlag, with a total of 300,000 copies in print by 1933.

In the year that Edlef Köppen joined the publishing house, Ivan Goll and Andrée Gide were signed up as authors, followed the next year by George Bernard Shaw, Max Herrmann-Neiße, and Oskar Loerke, and in 1922 by Emil Ludwig and Bertolt Brecht, whose *Baal* opened the series of Brecht publications. Of particular importance for Köppen was the acquisition of the cultural journal *Die Dichtung (The Poetry)*, which had previously been published by Roland-Verlag in Munich. A separate publishing house also named "Die Dichtung" was founded and affiliated with Kiepenheuer. After six months, Köppen became production manager of the department ("Verlag der Dichtung") and was primarily responsible for the publication of the journal.

For health reasons, Edlef Köppen was forced to leave Gustav Kiepenheuer Verlag on May 1, 1922. After a ten-week stay in two lung sanatoriums in the Taunus and in Upper Bavaria, he took over the management of the production department at Trowitzsch & Sohn publishers in Berlin in October 1922. This job also ended after a short time due to health problems. The work there, the artistic design of luxury editions, proved to be good preparation for what followed: In mid-May 1923, Edlef Köppen founded his own bibliophile publishing house, Hadern-Verlag, in Potsdam. The program included the publication of complete works, illustrated individual works, typographic specimen sheets, modern literature, and graphics. The ambitious project only produced a few works, in precious designs with woodcuts, etchings, and lithographs by contemporary artists, and was closed the following year.

The financial worries of the Köppen family—their only child, daughter Gabriele, was born in June 1924—ended in October 1925, when Edlef Köppen was appointed to the literary advisory board of the "Funkstunde" (Radio Hour) in Berlin, which was then the most ambitious and active of the German radio stations that were just starting up at the time. He helped shape the artistic program of the literary department, of which he had previously been a freelance contributor. In 1929, he was appointed head of the department. This was the job that finally fulfilled him completely, as he confessed in 1930, looking back: "The job began, bookseller, publisher, 'freelance writer.' Hardship began, hunger. I translated Heraclitus, wrote for daily newspapers, modestly filled the drawers with manuscripts. Finally, a profession came that gave more than the opportunity for menial labor: I now live in it. I love it. So, the work can begin."

Under Köppen, new forms of radio programming were developed, such as the "spoken oratorios" or the "improvised narration," in which contributors such as Alfred Döblin, Hermann Kasack, and Arnold Zweig deliberately did without manuscripts. Köppen was not only head of the literary department but also a contributor and director of programs. Unknown poets were offered the opportunity to reach the public through radio. In general, Köppen ushered in a new style in the relationship between the mass media and authors, which was due to his good contacts with literary Berlin. It is more than likely that Köppen also met Karl Kraus, the author of the monumental World War drama *The Last Days of Mankind*, by 1932 at the latest, when Kraus was realizing his twelve-part cycle of Jacques Offenbach's operettas at the "Funkstunde" (albeit for a different department).

Köppen's left-wing liberal, pacifist commitment as a writer, lecturer, and head of the literary department of the "Funkstunde" led to his suspension in April 1933 and, at the end of June 1933, to his dismissal without notice under the "Law for the Restoration of the Professional Civil Service." His publication of the *Army Report* (1930), his participation in events organized by the "League for Human Rights," and radio programs such as the pacifist radio play "Wir standen vor Verdun" (*We stood before Verdun*), which he broadcast in February 1931 on the occasion of the 15th anniversary of the Verdun offensive, had made him an unpopular figure among National Socialist party supporters. Köppen's appeal to the Reich Ministry for Public Enlightenment and Propaganda, to which the Berlin radio station had been subordinated on March 22, was rejected after half a year, as expected: "As head of the literary department of the Funkstunde GmbH. Berlin, Köppen had a decisive influence on the design of this station's predominantly culturally subversive program."

Köppen's successor at the "Funkstunde" was the right-wing writer Arnolt Bronnen. He himself was banned from publishing. His last independently published work under his own name was published by Cassirer Verlag in 1934: the light-hearted and harmless stories about building a house, *Vier Mauern und ein Dach (Four Walls and a Roof)*. He published other literary works, such as reviews and essays, under the pseudonym Joachim Felde, which he derived from his birth name Joachim Edlef Köppen. Through the mediation of a former radio colleague, Köppen was offered a job as head of the press department and press spokesman for a small film company in August 1934. Two years later, he joined Tobis Europa Film AG as head dramaturge. In this position, he was responsible for the scripts and worked with the film stars of his time on entertainment films (Jan Kiepura, Hans Albers, Willi Forst, Gustaf Gründgens, Emil Jannings, Paula Wessely, Karl Hartl, etc.). When Tobis came under the control of the Reich Ministry of Propaganda, Köppen came under increasing pressure. He refused to join the Nazi Party and to include pro-Nazi and anti-Semitic films in the program. This is evidence of the tightrope walk that Köppen had to perform: on the one hand, he was not a party member, his books were on the index, and he lived in a state of inner emigration; on the other hand, he was a member of the Reich Chamber of Literature until his death, took on a responsible position in film production, and had to work with film greats who cooperated with the Nazi state. He was no longer able to realize his plan of writing a book about the National Socialist state, for which he had been collecting documentary material since 1933. On February 21, 1939, Edlef Köppen died in Gießen of tuberculosis of the lungs and larynx. The obituaries remembered the chief dramaturge of Tobis and the author of *Four Walls and a Roof*, but made no mention of *Heeresbericht*.

Edlef Köppen's work on *Heeresbericht* coincided with his radio work. During the war, he had already decided to process his experiences in the form of a novel. Köppen obtained the historical documents, which the author had compellingly incorporated into his account of the experiences of war volunteer Adolf Reisiger, from his landlord, a senior official at the Army Archives in Potsdam. Other material, such as newspaper clippings and advertising cuttings, had already been collected by Köppen's mother during the last year of the war, especially those "not consistent with his accounts of the front."

The *Heeresbericht* was published in 1930, during the final phase of the Weimar Republic. It appeared in print through the Horen-Verlag, Berlin Grunewald, and a second edition was published in 1932 by Paul List Verlag, Leipzig, which had bought out the Horen-Verlag. It seems that the two editions did not include more than 10,000 copies, in addition to the English-language publication under the title *Higher Command* (1931). *Heeresbericht* could not assert itself in the flood of World War literature—112 new titles appeared in 1930 alone—a flood that characteristically only swelled about ten years after the end of the war, when the events began to appear comprehensible and processable, but also encouraged the politically exploitable formation of myths. In particular, the overwhelming competition from Erich Maria Remarque's *All Quiet on the Western Front* (1929) and Ludwig Renn's *War* (1928), with print runs of 3.5 million (Remarque) and 150,000 (Renn) copies respectively, left little room for the more literarily ambitious work of Köppen and his small publishing house. The broad spectrum of war books ranged from the communist left (Adam Scharrer: *Vaterlandslose Gesellen*, 1930) to the nationalist right (Hans Zöberlein: *Der Glaube an Deutschland*, 1931), and in addition to "democratic" war novels (Renn, Remarque, Plivier) also included a large number of nationalistic, folkish novels, which were also published in enormous numbers (Beumelburg, Schauwecker, Zöberlein). *Heeresbericht* nevertheless attracted a great deal of attention among connoisseurs. Gottfried Benn, Kurt Pinthus, Ernst Toller, and Kurt Tucholsky were enthusiastic about it. Oskar Loerke, poet, essayist, and editor at S. Fischer Verlag, confessed that of all the war books he had come across, *Heeresbericht* had made the greatest impression on him: "And this was the effect on me: I didn't read, I saw; I didn't see, I felt; I didn't feel, I experienced, spellbound and directly, an epic event of extraordinary diversity with cumulative, tragic force." Ernst Toller attested that Köppen, along with Plivier *(Des Kaisers Kuli*, 1930), had written the most courageous war book published in Germany to date, and wished him hundreds of thousands of readers in Germany and around the world. And Kurt Pinthus described the secret of Köppen's book: "Although it does not want to be fiction, but only a report, it nevertheless comes across as fiction."

The strong impression that reading the book makes even today is based on its authenticity, an authenticity that avoids both self-pity and sentimentality as well as the accumulation of horror. It is not only reinforced by the documents that are inserted between the descriptions of the experiences of the war volunteer, but also by the autobiographical traits that the author admits to in the blurb of the first edition: "I was born on March 1, 1893. Consequently, I was able to volunteer for the armed forces in August 1914, which I duly served from October 14 to October 18 in the West and East on His Majesty's Service as a gunner, private, corporal, vice-sergeant, deputy officer, reserve lieutenant. I did it with enthusiasm, with a sense of duty, with gritted teeth, with despair, until I was awarded the Iron Cross First Class and put in the madhouse."

With autobiographical foundation and stylistic precision, the development of the student war volunteer Adolf Reisiger is presented, who, due to his military virtues, undergoes an "honorable" career, at first rather reluctantly, from gunner to lieutenant of the reserve. His deployment in the war allows him to experience the cruel realities of everyday life on both the Western and Eastern fronts and to gradually see through the system of the war machine. The questions he asked himself at the beginning in October 1914 with irrepressible curiosity ("Where is the war?" "Are we at the front now?" "What is the enemy, who is lurking somewhere?" "How does this whole operation work?") are answered. The reader is vividly shown the development of the war from mobile warfare to the grueling war of position to the devastating material battles, especially as Reisiger's growing experience and promotion to squad leader broaden his perspective, transforming his lack of understanding and oversight into a disgusted realization of the senselessness of these actions. No one has been able to depict as vividly as Köppen what the completely new horror of gas attacks meant for the soldiers. Reisiger recognizes that war is murder by order, "the greatest of all crimes." Also because of the feeling of guilt for having participated for years and for having become complicit in the deaths of many people through his command, he resorts to insubordination in late summer of 1918 and is locked up in an insane asylum, finally in the fortress military hospital in Mainz, to await the end of the war.

Köppen's style is characterised by a sober and detached power of observation of laconic precision. Interwoven with numerous documents, which are intended to provide the most comprehensive overview possible of public consciousness and its manipulation, the fundamental contradiction becomes clear in the juxtaposition of war ideology and reality. The literary novelty of the montage technique was what made the book so appealing to many readers and critics. In early modern literature, montage techniques were initially used in poems and satirical sketches, especially in the context of Dadaism. Karl Kraus became a pioneer of the montage of temporal particles, both in the quotation technique of his magazine "Die Fackel" and, above all, in "The *Last Days of Mankind* " (1918/1922). Of course, knowledge of Alfred Döblin's *Berlin Alexanderplatz* (1929) (Köppen had worked with Döblin at the "Funkstunde") is also a prerequisite, as well as the earlier novels of John Dos Passos, who was widely read in Germany: *Manhattan Transfer*

(1925, German 1927) and, in particular, *Three Soldiers* (1921, German 1922), the first attempt to do justice to the events of the world war using advanced literary methods. Dos Passos and Döblin were the first to use the montage technique for the overall structure of the novel. Köppen does not go as far as Döblin in the seamless interweaving of the novel text with quoted particles of reality. Nevertheless, the interruption of the plot by the uncommented insertion of official decrees and orders, army reports, newspaper articles, and advertisements met with a great response from critics. For the first time, a German novelist had managed to apply the gestures of literary fashion to the subject of "world war," which had previously been attempted, independently of moral and political evaluation, with traditional narrative forms that dated back to the nineteenth century—this is still true for Remarque. Ernst Toller: "I know of no other war book in which words such as 'barrage', 'gas attack', 'shelter', and 'trench' are dissolved and pictorially re-created from the sensual-representational." Köppen achieved a similar expressiveness in his 1931 broadcast *We stood before Verdun*, in which he skillfully combined genuinely war-glorifying novel excerpts by, among others, Ernst Jünger *(In Stahlgewittern)*, Franz Schauwecker *(Aufbruch vor Verdun)*, Josef Magnus Wehner *(Sieben vor Verdun)*, and Werner Beumelburg *(Sperrfeuer um Deutschland)* with critical statements about the war, for example from letters from fallen front soldiers.

Even decades after its first publication, the outstanding literary and artistic quality of the *Heeresbericht* ensures a moving and exciting read. Köppen's major work is convincing in its power of expression as an appeal to conscience and a plea for pacifism. The First World War was his outstanding theme, which he accentuated and articulated differently in various creative phases. Remarque's enormous success was due to an unmistakable trivialization and sentimentalization, while Ludwig Renn's more modest work was due to its naive, mendacious attitude of mediation. *Heeresbericht* remains unachieved, unsurpassed as the most formally advanced war novel, carried by its own mixture of cool distance and passionate engagement, which at least comes close to Karl Kraus' *The Last Days of Mankind*. There is nothing more that can be said about this author and his book—perhaps just this (about Ernst Toller): "Hats off to Edlef Köppen."

Edlef Köppen and the Third Reich

Wilhelm Ziehr
Additions by Sven Dethlefs

Köppen's life in the Third Reich was affected in all three of his professions: as a director for the "Funkstunde" (Radio Hour), the literature program at Berlin radio, as the author of the *Heeresbericht*, and later as the head dramaturge of Tobis Europa Film AG.

Even before Hitler came to power in 1933, Köppen's role at the *Funkstunde* radio program began to be increasingly restricted. This process started in mid-1932, following the appointment of Franz von Papen as Reich Chancellor. Von Papen named Erich Scholz—an NSDAP member—as radio commissioner in the Reich Ministry of the Interior. The resulting personnel changes were officially described as a "radio reform," but in practice, they marked a political takeover. A so-called "monitoring committee," composed of party members, was tasked with overseeing these changes—an inherently absurd arrangement, as it meant evaluating their own party's growing influence. At the same time, new "cultural advisory boards" were established, but these bodies were far from democratic or open to artistic freedom.

By the end of 1932, Thomas and Heinrich Mann had already been removed from Köppen's list of proposed authors for the *Funkstunde* winter program. A prepared broadcast by Köppen titled *How They See Us*—a commentary on Germany from the perspective of foreign countries—immediately drew the attention of Radio Commissioner Erich Scholz and the Foreign Office. The political atmosphere was becoming increasingly tense; everything was now viewed with suspicion.

In early August 1932, broadcasting companies across the Reich had their operating licenses revoked. More damaging to the principle of free broadcasting was a statement made on October 28, 1932, by Ministerial Counselor Dr. Strunden of the Prussian State Ministry. He declared that, under the new political circumstances, the era of independent programming by *Funkstunde*'s department heads was over. A new official position—Program Director—was introduced, and filled by Richard Kolb, a member of the National Socialist Party.

It was also no longer possible to observe Bredow's written *constitution of broadcasting*: "The broadcasting service does not serve any party. Its entire news and lecture service must therefore be strictly non-partisan." Köppen was removed as head of the literary department. He also lost oversight of the radio plays, a responsibility recently transferred to him by the director before his dismissal. His successor was Arnolt Bronnen, whom Axel Eggebrecht later described as "Goebbels' pioneer."

Köppen tried in vain to resist the reduction of his authority, which violated the terms of his employment contract. His dismissal became inevitable after Hitler's rise to power in January 1933. Although his contract was set to expire in June 1934, he had been terminated well before that date and was barred from entering the Berlin radio station. His removal was officially justified under the Second Implementation Decree of the Professional Civil Servants Act. He was labeled "an exponent of the opposing view to National Socialism" and deemed to lack "the reliability required by Section 4" of the decree.

Although Köppen's *Heeresbericht* met with moderate success after its initial publication in 1930, a second edition was released in 1932 by List Verlag in Leipzig, which had acquired the original publisher, Horen Verlag (Berlin). However, the book had little chance of reaching a broader audience after Hitler came to power. Almost immediately, the Nazi regime began censoring literature. Berlin librarian Dr. Wolfgang Herrmann compiled a "blacklist" of books to be removed from bookstores and libraries. On May 1, 1933, he passed these lists to the National Socialist German Student Union to guide the looting of libraries, lending institutions, and scientific collections as part of the "Campaign against the Un-German Spirit." The exclusion lists were widely published in newspapers and magazines in early May. The campaign had been officially launched by the German Student Union on April 12, 1933, and culminated in public book burnings on May 12. Works by authors such as Einstein, Freud, Thomas and Heinrich Mann, Remarque, Schnitzler, Kafka, Heine, Hesse, and Marx were among those destroyed in these acts of symbolic "cleansing." In a 1977 interview with *Deutschlandfunk* editor German Werth, Köppen's wife, Hete Köppen, stated that *Heeresbericht* was also among the books burned during these events.

Despite such a public condemnation, Köppen remained a member of the Reichsschrifttumskammer (Reich Chamber of Literature, founded in November 1933). This organization was part of the Reichskulturkammer (Reich Chamber of Culture), a government agency established on September 22, 1933, by Reich Minister for Public Enlightenment and Propaganda, Dr. Joseph Goebbels, as a professional organization of all German creative artists. Its purpose was the removal of Jews from German cultural life, and the control of any cultural activity with the aim to promote Nazi propaganda. Membership required an Aryan certificate, and denial of membership practically resulted in the inability to find employment.

In 1935, *Heeresbericht* was officially banned, and remaining copies were confiscated by order of the Leipzig police president. By 1938, the book had been added to the regime's *List of Unwanted and Harmful Literature*. Earlier, on January 27, 1934, Köppen had contacted List Verlag in Leipzig to submit the manuscript for a new and politically harmless work titled *Vier Mauern und ein Dach* (*Four Walls and a Roof*). In response, the publisher conducted "inquiries" into Köppen's political reliability—specifically, his stance in relation to National Socialism. According to a report from a publishing representative, there was at that point no clear indication that Köppen was being actively persecuted or that his work was officially blacklisted: "Incidentally, I also made confidential inquiries about any possible incriminations of your name on this occasion. On the basis of this, and according to authentic assurances, I am able to tell you that nothing is being held against you, but that you may claim all the rights of a free German writer within the Reich Association of German Writers, the Reich Chamber of Literature and the Reich Office for the Promotion of German Literature. The suppression of your *Heeresbericht* thus probably only bears the accusation of insufficient heroic attitude."

In 1936, Köppen got a position as dramaturge at Tobis Europa Film AG, a Berlin film company with production studios in Johannisthal, in the southeast of the capital. Köppen had established relations with the film industry before he was forced to leave the radio, and he had already found contact with Tobis in 1930. He had also written several film scripts. Tobis was one of the four major German film companies along with Terra Film, Bavaria Film, and the famous UFA with its stars Marlene Dietrich and Fritz Lang. Goebbels heavily influenced and controlled the German film industry, transforming it into a tool for Nazi ideology and propaganda. The guiding principle of his policy was: "I regard the establishment of the new ministry ... as a revolutionary act of the government, since the new government no longer intends to leave the people to their own devices."

Köppen was summoned to talks by Goebbels in 1933 (an exact date cannot be determined), and on February 17, 1937.

The content of his conversation with Goebbels remains uncertain. In a letter to his wife dated February 14, 1937, with several pages and later additions up to February 18, 1937, one can feel Köppen's mood at the time and sense his grave concern, which becomes tangible in the following lines:

"It's as if I were bewitched ... I can't move! This week I have been ordered to Dr. Goebbels, but the day is still unspecified ... you can't ask ... I can't leave. Please don't lose your nerve ... I have to keep it too! If the talk with the minister goes well ... I'll take a few days' vacation and come ... Otherwise, don't worry ... I'm not bad ... and I don't have time for loneliness and sadness, thank God. Phew ... just get this over with first. Going with Jannings, Gründgens and Willi Forst ... it's a bit strange ... like graduation exams at high school ... our little Kö in the big wide world ... Had a weekend in Saarow (at Gründgens'), worked very well."

On the 18th, he wrote the liberating addition: "Dear Ones! Dr. Goebbels settled. Case settled." He had good reason to be deeply concerned due to his refusal to follow Nazi ideology, and because as early as 1933, his former superiors at the *Funkstunde* were put in "protective custody" in the Oranienburg concentration camp, including the director Hans Flesch.

As chief film dramaturge at Tobis, Köppen held broad responsibilities and enjoyed significant autonomy in negotiations and collaborations with major film figures of the time. He worked closely with leading actors such as Emil Jannings and Gustav Gründgens, two of the biggest stars in German cinema. He also traveled to Poland to see the tenor Jan Kiepura, one of the most popular singers of the time, and then to the actress Paula Wessely, who Goebbels wanted to hire from Vienna to Berlin. Goebbels allowed popular authors like Erich Kästner to work in the film industry, under a pseudonym, and to write the script for the film "Münchhausen" with Hans Albers in the title role, although Kästner's books had been burned in 1933. Köppen continued to publish too, sometimes under pseudonyms. In 1939—a time when the *Heeresbericht* had already been publicly condemned—a second edition of his *Four Walls and a Roof* was released by Drei Masken Verlag in Berlin.

Edlef Köppen died on February 21, 1939, in Gießen, Hesse. He passed away in a lung sanatorium due to complications from his World War I injuries.

Letters

Edlef Köppen wrote letters from the front to his confidante Maria Fellenberg, the girlfriend and later wife of Hermann Kasack (his lifelong friend and later colleague at the Gustav Kiepenheur Verlag in Potsdam).

As long as the feeling of having volunteered for military service lasted, the echo of the war seemed rather incidental or followed the familiar tone of enthusiasm and the spirit of optimism in August 1914.

On August 31, 1914, just before his transport to the front, Köppen wrote to Maria Fellenberg from Burg (near Magdeburg), where he was undergoing training:

"My joy is great, enormous! Despite the setbacks that will surely come more often, despite the many serious, grieving faces, I go out joyfully. It is a feeling, so uplifting, so sacred, as I never had it: With its banners, with pure German conscience into the crowds of our miserable enemies! No matter how numerous they are. They will all have to recognize that Germans cannot be crushed. Wherever hundreds fall and bleed to death, thousands will rise again.

A holy war! And ... I believe this with all the firmness of a rock ... a holy victory!"

Almost three months later, on November 17, 1914, when his unit was east of Arras, his mood had not changed:

I'm always fine here. And there's not enough work. If only I could read here and work. It's not possible here. There's hardly a newspaper in which there is anything sensible to read. But I don't complain. I only give gratitude. And I am completely full of confidence. Whatever happens, happens. Every evening and every morning my heart is full of thanks ...

Your happy Edlef

On December 3, 1914, Köppen dispels the notion that people at home have *"many terrible ideas about life here."* Yet the war has not left him as untouched:

"Despite all the grief and the often unspeakable horrors here, everything is pure happiness for me! We will talk about it when we all see each other again."

Twenty days later, his letter takes on a more serious tone:

"In the last few days I have so often stood at the gates of death with burning eyes ... many comrades around me fell ... torn apart by enemy grenades ... I always remained unharmed ... I will continue to be mercifully protected ..."

In January 1915:

"Longing ... burning longing from day to day. It flares up in me everywhere, screams through the nights ... sometimes cries through the days, when a moment arises to be quiet and listen, in the midst of all the agony and killing and the boundless, incomprehensible sadness that every new hour of war must bring us. Dear God, if only it would all be over soon!"

On January 6/7, 1915:

"I am writing these lines on guard duty. Without sleep for forty-eight hours. We all tremble in expectation of an alarm that could come at any hour! ... and calls us into the unknown of battle."

On January 27, 1915:

"Three days of constant shooting, with our and the enemy's guns going off in a constant clash, slowly makes everyone's hearts tremble. Only afterwards do you feel—how close you are to dying here all the time ... But you already feel something like a scream from deep within your soul, something that sounds above everything: ... why can't you live!!!!!!"

On February 2, 1915:

"There are many things here that are terribly tormenting. Sometimes at night I think . . . During the day, you sit lurking at the gun and see one shell after another roar in front of us . . . First far, 100, 150 meters . . . Then closer and closer . . . 50 meters . . . 25 meter . . . 15 meters . . . and then, then . . . then your eyes suddenly open . . . there, there is death . . . 5 meters. . . . Well . . . well, . . . the next one will hit you . . . 5 meters . . . it has to come . . . parents . . . brother, . . . relatives . . . all dear friends . . . It's all over in a minute . . . oh! just seconds . . . Then everything is clear again . . . then the eyes light up . . . Then you listen for the command . . . 'Fire' . . . and pull your own gun . . . everything trembles. . . . And then you smile. Make a joke. . . . Everything is forgotten . . . Then your heart beats so fast, Oh happiness of being a soldier!!! Death . . . You can't scare me . . . You, Death!! Brother!! Not true . . . You are good? Not true. You are . . . also an angel of our God . . . It makes you weak. Not fear!! I'm not afraid of anything anymore. But it makes you weak. This back and forth between life and death . . ."

"My thoughts, my ability to think, is also so inferior now . . . Heaven, if only the war would come to an end!!!"

The experiences with the suffering and dying men at his side or the hours spent in mortal fear did not prevent him from falling back into old phrases. On July 12, 1915, he wrote to Maria Fellenberg to console her for the suffering of her seriously ill mother:

"Force yourself inwardly to be a German girl. One who is capable of any sacrifice. Think of the many who sacrifice the dearest thing they have for their Fatherland every day . . . without war . . . and without weakness. That is German! And that is good!"

At the time, Köppen still sought refuge in faith, as he wrote on the evening of July 26, 1915:

"How bad I felt a year and a half ago. I had completely lost faith in God. A thousand doubts tormented me. I found no trust in a being that is supposed to rule us all. Only the war made everything different. When I saw the first dead . . . I still remember it like today . . . suddenly an insight

came to me. And when the first time a grenade tore apart eight comrades in front of my eyes, who were busy loading Christmas packages . . . then I thought about it . . . And then I slowly learned to believe."

When Köppen was awarded the Iron Cross in October, his convictions blossomed once more—on October 26, 1915, he wrote:

"Of course I will always be sensible and careful. But the Iron Cross alone, which I have been allowed to wear for days, obliges me to fulfill my duty to the end, wherever I am posted.
I will not be spared because they know I can accomplish something. And I am grateful for that."

In his letters to his confidante, he expresses how deeply he was affected by the death of his and Kasack's friend, Gerhard Lepsius. However, he avoids burdening her with details about the difficult moments in which he wrote poems for *Die Aktion* or composed the desperate verses of 'Loretto I' for Hermann Kasack. At times, though, his emotional strain becomes apparent—such as in August 1916, when he admits that even during leave, things are no longer the same. *"Perhaps it is only later that one fully realizes how the war completely destroys one! I can hardly read for five minutes . . . then I'm done. I don't feel like writing . . . sometimes I don't feel like going out, even though I feel oppressed and insecure alone in my room. I want to sleep . . . always sleep. Everything seems old to me . . . a thousand years old . . ."*

Hermann Kasack confirms this image of Köppen and his mental breakdown. On September 8, 1916, Kasack wrote about an encounter with Köppen:

"Edlef is so finished! In some moments, hours, it's downright frightening. He can only stand the service with great difficulty. Besides, his nerves . . . one is so helpless . . . towards him."

As early as the beginning of 1917, in a letter to Maria dated January 15, Köppen reflects on his future and resuming his studies in Munich:

"There is no sense in complaining. It is bitterly clear to me that I gain nothing, nothing from the war... that in peace I simply go back to school and start all over again. And whether it is possible to preach school wisdom to a person who has experienced the ultimate in horror and death and transience in the field and has thus matured, one will just have to be left to time."

In August and September 1917, Köppen had to realize that the war would not end soon and that he would have to *"still vegetate out here."* And on April 29, 1918, alluding to the military situation, he explained hopelessly:

"We have boundless difficulties behind us. Day after day, man..., into the unknown... starving... freezing... always riding and riding. In the evening, laid out on the cold ground or in wet straw... no mail. No contact with home... oh... I have suffered as I rarely did in these terrible years."

The descriptions also include the destroyed environment, the shot-up villages, the fields churned up by shells, as on August 2, 1918:

"Oh... if it weren't all so terrible. Just looking at the landscape here... these villages trampled, simply crushed in the vilest, most brutal way... you think you can't bear any more. And then all the dead... all the misery."

On September 26, 1918, after the failed offensive in the West, he writes:

"Attack... I saved nothing but my life and the few things I happened to be wearing. Maria... five minutes longer... and I would now be a prisoner somewhere... I don't want to talk about it. Can you understand what it means to have to run between your own and enemy machine guns... that you are then at the end of your wits and your strength!"

Finally, on November 21, 1918, Köppen describes the end of the war:

"You know, it's something tremendous, an unspeakable feeling... LIFE ... After four years of death, finally! It is as if a new incarnation has taken place. I sit at home in my room, with my books, at my desk... and no one is giving orders. And no one is calling. I do what I want. I walk the streets. Or into the forest... and no one chases after me. No one shoots. Nowhere do the dead lie or broken people weep..."

Wilhelm Ziehr. *Edlef Köppen und die literarische Darstellung des Ersten Weltkrieges.* Lecture in Genthin, September 2013

Professor Jens Malte Fischer studied German language and literature, history, and musicology, as well as singing, in Saarbrücken, Frankfurt am Main and Munich. He was a professor of modern German, comparative and general literary studies at the University of Siegen from 1982 to 1989 and a professor of theatre studies at the Ludwig Maximilian University of Munich from 1989 to 2009. He was a fellow of the Wissenschaftskolleg Berlin and is a full member of the Bavarian Academy of Fine Arts (Munich), the Academy of Sciences and Literature in Mainz, and the German Academy for Language and Literature. In 2014, he received the Bavarian Maximilian Order for Science and Art. He lives in Munich.

Dr. Wilhelm Ziehr studied history, Romanistik, German literature, and art history in Tübingen and Paris, and graduated with a PhD in Eastern European history. He worked as editor at Brockhaus Encyclopedia, Wiesbaden, as editor-in-chief of the book series *Weltreise* in Lucerne, and as a freelance author and editor. In 1987, he became the co-founder of the publishing house Schweizer Lexikon, Mengis und Ziehr, Lucerne, editor and editor-in-chief of the *Schweizer Lexikon*. He lives in Wilhelmshorst/Brandenburg.

Dr. Sven Dethlefs studied biochemistry in Tübingen, Berlin, and Milan, and completed a PhD at the Pasteur Institute Paris. He was a partner at McKinsey & Company and works in the pharmaceutical industry. He lives in Philadelphia/USA.

Acknowledgments:

Professor Jens Malte Fischer for kindly allowing the use of his afterword from the DVA Munich edition of the *Heeresbericht*.

Dr. Wilhelm Ziehr for his extensive research on Edlef Köppen and for generously providing his study materials and publications.

Colonel Jarrod Stoutenborough from the United States Marine Corps, Marine Artillery Detachment, Fort Sill, for reading a draft and providing comments on military rank translations.

Leni Dethlefs and Anne Kathrin Wille for providing many beautiful cover design ideas.

DeepL.com for its thoughtful adaptation of very long German sentences to modern English.

ChatGPT for finding translations to songs and poems that required creative interpretation and a more artistic translation.

The team at ebooklaunch.com for the cover design, editing, and proofreading services.

Andrea Reider for formatting services.

Allworldwars.com for a dictionary of military expressions and ranks.

Project Gutenberg for an online publication of the Heeresbericht.

Text sources: DVA Munich, 2nd edition 2004 and project-gutenberg.org

www.ingramcontent.com/pod-product-compliance
Lightning Source LLC
LaVergne TN
LVHW041742060526
838201LV00046B/879